PHENOMENOLOGICAL ASPECTS OF WITTGENSTEIN'S PHILOSOPHY

SYNTHESE LIBRARY

STUDIES IN EPISTEMOLOGY,

LOGIC, METHODOLOGY, AND PHILOSOPHY OF SCIENCE

VOLUME 268

BYONG-CHUL PARK
Pusan University of Foreign Studies, Pusan, Korea

PHENOMENOLOGICAL ASPECTS OF WITTGENSTEIN'S PHILOSOPHY

SPRINGER-SCIENCE+BUSINESS MEDIA, B.V.

A C.I.P. Catalogue record for this book is available from the Library of Congress.

ISBN 978-94-010-6165-0 ISBN 978-94-011-5151-1 (eBook)
DOI 10.1007/978-94-011-5151-1

Printed on acid-free paper

Originally published by Kluwer Academic Publishers in 1998
Softcover reprint of the hardcover 1st edition 1998

TABLE OF CONTENTS

PREFACE

In his writings around 1930, Wittgenstein relates his philosophy in different ways to the idea of phenomenology. He indicates that his main philosophical project had earlier been the construction of a purely phenomenological language, and even after having given up this project he believed that "the world we live in is the world of sense-data,"[1] that is, of phenomenological objects. However, a problem is posed by the fact that he does not appear ever to have given a full, explicit account of what he means by his 'phenomenology', 'phenomenological language', or 'phenomenological problems'. In this book, I have tried to unravel the nature of Wittgenstein's phenomenology and to examine its importance for his entire work in philosophy.

Phenomenology can be characterized as philosophy whose primary concern is what is immediately given in one's experience. This 'immediately given' is not merely impressions inside one's mind, but includes also the part of objective reality that impinges upon one's consciousness. Thus, an aim of phenomenological enterprise is to grasp this objective reality by attending to immediate experience. Husserl's phenomenology is in fact a case in point.

Similarly, Wittgenstein's main attention throughout his entire philosophical career is directed to one's immediate experience. What makes Wittgenstein's phenomenology unusual, however, is the way he views immediate experience. For Wittgenstein, especially the early Wittgenstein, the nature of the immediately given is unique in the sense that the possibility of logic is also based on it. Indeed, the fact that Wittgenstein sees the foundation of logic in immediate experience is what makes his philosophy a kind of phenomenology. In the *Tractatus*, all logical forms are combinations of logical forms of immediately given objects of experience. Also, immediate experience can only be grasped and understood by means of language, because Wittgenstein views language as the ultimate medium through which we understand the world. Therefore, the important problem for him is the way we use language to describe immediate experience. In my study, I examine Wittgenstein's concern with immediate experience and the way we describe it, a concern which guides his philosophical journey through the phenomenological problems that pervade his work.

Chapter one discusses the general problems with regard to Wittgenstein's phenomenology. I widely examine and discuss the philosophical background of Wittgenstein's phenomenology, which should be clearly distinguished from any form of phenomenalism. In order to clarify what Wittgenstein means by his phenomenology, I not only examine the way Wittgenstein uses the words 'phenomenology' and 'phenomenological' in his own writings and other related materials but try to trace the possible influence from which Wittgenstein might have received the usage and the idea of phenomenology. Among many possible influences are those of theoretical physicist Ludwig Boltzmann as well as Heinrich Hertz and Ernst Mach.

Chapter two consists of a detailed discussion of Wittgenstein's phenomenology in the *Tractatus*, the focus being on how Wittgenstein develops his phenomenological ideas, which consist of the Tractarian logic and logical analysis, and on how he comes to modify his earlier view toward the middle period. Most importantly, I discuss Russell's influence on the development of Wittgenstein's early philosophy, and I carefully examine the philosophical interactions between Russell and Wittgenstein with regard to Russell's theory of acquaintance, for Wittgenstein's reaction to Russell's theory results in Wittgenstein's phenomenology. In addition, I discuss the problem of solipsism in relation to the general characteristic of phenomenology, bringing the ideas of Husserl, Russell, and Wittgenstein into interesting comparison, as solipsism occurs in each of these philosophers' epistemological concerns.

Wittgenstein's middle-period writings show that Wittgenstein is in transition. Chapter three discusses how he not only rejects some of the most important doctrines of the *Tractatus* but adopts new logic that eventually grows into the notion of grammar, and how he shifts language paradigms, namely from the phenomenological language of the *Tractatus* to the physicalistic language of ordinary use. In light of this shift, I examine his middle-period treatment of solipsism and the problem of phenomenology of time. The change in the language paradigms apparently makes it impossible for Wittgenstein to conduct phenomenological enterprise. In fact, the abandonment of earlier logic, which is phenomenological in nature, waters down the phenomenological aspects of Wittgenstein's philosophy from the middle period on to the extent that he declares "there is no such thing as phenomenology, but there are indeed phenomenological problems."[2] I examine how Wittgenstein continues to wrestle with these phenomenological problems as he develops the language-game idea.

In chapter four I discuss Wittgenstein's treatment of color concepts. The problem of color is very phenomenological for Wittgenstein in the sense that the color he examines is color we immediately experience. I show that Wittgenstein tackles this phenomenological problem of color from the early period through the very last. In the early period colors are the example of Tractarian objects, so that from the beginning, Wittgenstein has no interest in the theory of physical color, a point well presented and supported in his *Philosophical Remarks*, where he repeatedly says of the theory of color that contains no hypothetical elements. As the language-game idea develops, Wittgenstein directs his concern toward how we can express and describe the color of immediate experience in our language-game, and so I conclude the color chapter with a discussion of the language-game of color.

In chapter five I develop the idea that Wittgenstein's use of the word 'aspect' means a phenomenological object. The main purpose of Wittgenstein's frequent use of the word 'aspect' along with 'aspect-seeing' and 'aspect-change' in his late writings is not to entertain Gestalt psychology or philosophy of psychology *per se*, but deeply related to his earlier ideas of phenomenology. Examining his use of the word in middle and late period writings, I show that Wittgenstein's 'aspect' does not have its familiar meaning of 'side' or 'facet'. For Wittgenstein, the word 'aspect' means a way of appearing, a phenomenological object picked out by logically proper names like 'this' and 'that'.

I conclude my study of Wittgenstein's phenomenology with criticism of other phenomenological interpretations presented since the 1950s. Some intend to identify Wittgenstein's phenomenology with Husserlian phenomenology, while others dismiss any claims that Wittgenstein was a phenomenologist. Also, some interpret that Wittgenstein's early and early middle-period works are prone to phenomenalism. As I show in chapter one, Wittgenstein's philosophy cannot be identified with phenomenalism in any sense, as Russell's philosophy in the early 1910s cannot be. I also criticize the claim that Wittgenstein was never a phenomenologist as we have reasonable ground to assert that Wittgenstein indeed put forward what can be characterized as the main concerns of phenomenology. We have to be cautious in this regard of simplification of the phenomenological aspects appearing in Wittgenstein's philosophy. Ignoring the context and background of the problems out of which Wittgenstein's phenomenology grew, some philosophers attempt to identify Wittgenstein's phenomenology as belonging to the so-called phenomenological movement. But in my evaluation of this attempt I propose that Wittgenstein's phenomenology is a different kind from Husserlian phenomenology.

This book is a revision of my doctoral dissertation, which I completed in 1995 at Boston University under the supervision of Professor Jaakko Hintikka. One of the purposes in my study is to continue and develop in further details what Professor Hintikka set out in his phenomenological interpretation,[3] and to present the phenomenological aspects in different stages of Wittgenstein's philosophy in a systematic manner.

I wish to express my gratitude to Professor Hintikka for his philosophical stimulation from the conception of the topic to the completion of this book. I also wish to thank Professors Judson Webb and Juliet Floyd of Boston University who read drafts of manuscript and provided penetrating comments for feedback. While I was writing chapter 5, I have benefited from the Wittgenstein Archives at the University of Bergen. I thank Claus Huitfeldt and

Maria Sollohub of the Wittgenstein Archives for kindly providing me with very helpful data from computer search. I would like to gratefully acknowledge H.K. Choi for sacrificing his time to make my incomplete English more tolerable. Finally, I would like to thank my parents. Without their encouragement and support, the completion of this book would not have been possible.

CHAPTER I

WITTGENSTEIN'S PHENOMENOLOGY

1. WITTGENSTEIN'S PHENOMENOLOGY

According to his surviving works, published or otherwise, it is evident that throughout his entire philosophical life Wittgenstein's philosophical attention is directed to immediate experience. Seemingly, Wittgenstein is concerned with empirical knowledge and its foundations for which his mentor Russell cared much. But Wittgenstein's problem is much more complicated and interesting than just epistemological grounding, for at the heart of the matter it involves what we may properly call phenomenology. It is not entirely surprising to see that Wittgenstein actually deals with phenomenological problems if we carefully review and examine his writings. Most notably, Wittgenstein himself uses the term 'phenomenology' or 'phenomenological' in his writings after 1929 when he came back to professional philosophy at Cambridge. The first four notebooks of 1929-30,[1] the *Philosophical Remarks*,[2] and the Big Typescript[3] are the chief sources in which Wittgenstein's own references to phenomenology occur most frequently. In particular, the Big Typescript contains a chapter entitled "Phenomenology," which begins with a section title "Phenomenology is Grammar." This simply cannot be viewed as a special case of, or a digression from his entire philosophical context. We even have a report of Wittgenstein declaring in his own words "You could say of my work that it is 'phenomenology'."[4] We also have testimony by Rush Rhees indicating that Wittgenstein once planned to establish a purely phenomenological theory of color.[5] These records of informal conversations and discussions date from 1929-1930 and are consistent with what Wittgenstein actually wrote during that period.

In fact, the *Philosophical Remarks* begins with the problems that Wittgenstein calls phenomenology and phenomenological language. Throughout his notebooks of 1929-1930, his philosophical struggles and frustrations with regard to the possibility of phenomenological language lead him to a settled philosophical pronouncement: the rejection of phenomenological language as a primitive language. However, in addition to this reassurance that phenomenological language is impossible, a conclusion reached in November 1929, Wittgenstein in the *Philosophical Remarks* strongly suggests that there are phenomenological problems. Indeed, what Rush Rhees testifies as Wittgenstein's attempted phenomenological theory of color is suggested in the *Philosophical Remarks* in several places, where Wittgenstein makes the point that he does not want to establish anything like a physical color theory, but "a psychological or rather phenomenological color theory. It must be a theory in *pure* phenomenology in which mention is only made of what is actually perceptible and in which no hypothetical objects occur."[6]

This statement reveals more or less what Wittgenstein means by his phenomenology, namely that he is concerned with the 'color' we immediately experience and with the mention of it in a hypothesis-free manner. Usually, talking of phenomenology instantly reminds one of Husserl and his unique philosophical method of phenomenological epoché, an essential doctrine of his phenomenology. Thus we might expect Wittgenstein also to have some sort of a methodological principle equivalent to the Husserlian epoché. It is clear that Wittgenstein's interest is directed to what is immediately given in one's experience, a point not much different from where Husserl's phenomenology starts. However, unlike Husserl, there is no mention of a methodological principle similar to epoché in Wittgenstein's phenomenology. Perhaps Wittgenstein had no need for a special method in his phenomenological enterprise. Wittgenstein's discussion of phenomenology usually coincides with the problems of phenomenological language, i.e., the way we describe immediate experience. Thus, in the case of his attempted phenomenological color theory above, Wittgenstein is concerned with 'the color we immediately experience' (the content of immediate experience) and 'the mention of it in a hypothesis-free manner' (the way in which we describe it). So we can say that Wittgenstein's phenomenological problem is the problem of immediate experience and the language that describes it. Whether he was familiar with Husserl's works or not, Wittgenstein did not seem to have a general interest in the transcendental aspects of phenomenology in the Husserlian sense. This can be further supported by Wittgenstein's attitude against the synthetic *a priori* nature of Husserlian phenomenological statements.[7]

There have been intermittent attempts to compare Wittgenstein's philosophy with Husserlian phenomenology or to relate it to the context of the phenomenological movement. But as I will discuss the early attempts later in chapter 6, they seem to fall short of grasping the full scope of Wittgenstein's phenomenology, partly because they fail to see the evident phenomenological feature immanent in the logical analysis of the *Tractatus*, and partly because they overlook phenomenological problems pervade many different stages of Wittgenstein's work. The facts that Wittgenstein's references to 'phenomenology' and 'phenomenological language' mostly appear in his middle-period writings, and that he declares the impossibility of phenomenological language should not lead one to think that his phenomenology is confined to a certain period of time in his philosophy. It is true that the *Philosophical Remarks* gives the impression that phenomenology is impossible because phenomenological language is impossible, as Wittgenstein states in the first section of the book.[8] What may be true is to say that Wittgenstein's use of the term 'phenomenology' or 'phenomenological language' ceases to appear after appearing frequently for a certain period of time. But that neither implies that his earlier philosophy was not phenomenological, nor that he turned away from what he believed to be the existing phenomenological problems. In no way for Wittgenstein does the rejection of phenomenological language also mean the rejection of phenomenological problems. For him, this rejection only means the replacement of phenomenological language by physicalistic language, a transition in language paradigms after he realizes the true nature of our language which belongs to the physical system.[9] That is why he could set a new goal after denying the possibility of phenomenological language in the first section of *the Philosophical Remarks*. It is to show what is essential to the class of languages which serve their purpose, and to an immediate representation of immediate experience.

Even though Wittgenstein no longer believes in phenomenological language, he indeed keeps addressing phenomenological problems by directing his attention to immediate experience and the task of hypothesis-free description of immediate experience. But as the pure description of immediate experience without phenomenological language is not possible, he opts to maintain that his phenomenology would be the grammar of the description. This point is not only presented in the first section of the *Philosophical Remarks* but is also supported by one of his middle-period slogans "Phenomenology is grammar," which appears in the Big Typescript.

However, that Wittgenstein's phenomenology is not a by-product of a short-lived whim is decisively shown in his very last writings. Whether he believes in phenomenological language or not, Wittgenstein has phenomenological problems in mind and keeps trying to cope with them up to the very last days of his life. In 1951 he asserted,

> There is no such thing as phenomenology, but there are indeed phenomenological problems. (*RC*, I, sec. 53; cf. III, sec. 248)

What would possibly be the phenomenological problems in question? At this late date Wittgenstein tackles color concepts once again. In his middle period, he wanted to establish 'a phenomenological color theory where no hypothetical objects occur.' It was the idea of the same period when he rejected the possibility and necessity of phenomenological language. Almost 20 years later and on the verge of death, Wittgenstein tries to deal with this matter once again but without phenomenological language. This is nothing other than a mature presentation of what he set out to do after the rejection of phenomenological language in the first section of the *Philosophical Remarks*. In it, he states that he will do away with phenomenological language yet find a way to do the same job that phenomenological language would have done.

A recognition of what is essential and what inessential in our language if it is to represent, a recognition of which parts of our language are wheels turning idly, amounts to the construction of a phenomenological language. (*PR*, sec. 1)

This statement clearly reveals the nature of Wittgenstein's phenomenology, which cannot be successfully discussed without connecting its problems to language. In a way he secures a place for addressing phenomenological problems within the physicalistic language system by attempting to figure out the essential grammar of our language. If he is successful, he really has no need to postulate such a thing as phenomenological language, in which he once believed.

There is a further justification enabling us to characterize Wittgenstein's phenomenology this way. It is his middle-period use of paired terms 'first system' and 'second system', or 'primary language' and 'secondary language' with regard to the two language systems. According to Wittgenstein's use of these terms, it can be noted that by 'first system' or 'primary language' he means phenomenological language that describes immediate experience, whereas by 'second system' or 'secondary language' he means physicalistic language.[10] Primary language is a language that serves to describe phenomena as they appear without any hypotheses. It mainly describes what is immediate; it does no more than pure description. On the other hand, secondary language is one that we employ to explain and predict by using hypotheses and by establishing laws of physics. Its function is more than description. Needless to say, Wittgenstein's phenomenology is closely related to the first system or primary language. And that is one way that Wittgenstein characterizes his phenomenological problems. It is once again the problem of immediate experience and the language that describes it.

If Wittgenstein's phenomenology is undoubtedly related to what he calls phenomenological language, his distinction between phenomenological language and physicalistic language and his shift of language paradigms in favor of physicalistic language must be very important in understanding his phenomenology. Although I will discuss the shift of language paradigms in chapter 3, a brief look at his distinction between two languages will be helpful. One thing we should be careful about is what exactly Wittgenstein means by phenomenological language and physicalistic language.

If phenomenological language is a language that describes what is given in immediate experience, what is physicalistic language? Although Wittgenstein relates physicalistic language (or secondary language) to the system of physics, it does not mean that his physicalistic language is *the* language of physics. It may include the propositions of physics.[11] But his term 'physicalistic language' is not exclusively used to refer to them, for Wittgenstein sometimes means everyday ordinary language by his term 'physicalistic language'.[12] Wittgenstein's physicalistic language is best understood when we compare it to his phenomenological language. Because phenomenological language describes what is directly given, the references of its basic symbols are immediate sense-data. As will be discussed later in chapter 2, sense-data are the most immediate data given in our experience. Since no hypothetical elements are added, we often describe them by 'this' or 'that'. On the other hand, physicalistic language is a language which adds hypothetical elements to the given data in representing them. Thus, the references of its basic symbols are physical objects, and we often use the words 'book' or 'table' to refer to those objects. In this sense, Wittgenstein's physicalistic language does not mean the same thing as the language of physics.[13]

Another thing to examine before we advance is where possibly Wittgenstein obtained paired terms such as 'first system' and 'second system' or 'primary language' and 'secondary language' which mean phenomenological language and physicalistic language respectively. Interestingly, Rudolf Carnap was using the distinction of 'primary' and 'secondary' in his writings. In his 1928 book, popularly known as the *Aufbau*, Carnap regards what is given in our direct experience as 'primary' and what is an abstraction from the given as 'secondary'.[14] In this period, Carnap's concern seems to be directed to what is epistemically primary in logical construction, and thus he emphasizes the given, which is characterized as direct experience itself. But in a 1932 paper that appeared in *Erkenntnis*, Carnap uses the term 'primary language (*die erste Sprache*)' as a more appropriate reference to his protocol-language.[15] For Carnap, the protocol-language deals with the given. It is 'language of direct experience' or 'phenomenal language' which seems similar to Wittgenstein's primary language or phenomenological language. The simplest statements in Carnap's protocol-language also refer to the given and describe directly-given experience. Although it is remarkable that Carnap also used the term 'primary language' to describe the given, it is not clear in fact Carnap's use of the term influenced Wittgenstein.[16]

The idea of 'primary' and 'secondary' also appear in Frank Ramsey's writings. In his "Theories"(1929), Ramsey talks about the primary system in a way that instantly reminds us of some ideas from the *Tractatus*:

> It [a notation of the primary system] might in the first case consist of names of different types any two or more of which conjoined together gave an atomic proposition . . .[17]

This is very similar to how Wittgenstein might talk about his Tractarian ideas of an elementary proposition, which consists of names of different logical types. In fact, Ramsey was so influenced by the *Tractatus* that he visited Wittgenstein several times in the 20s while the latter was a schoolmaster, not only to discuss philosophical problems but to persuade him to return to academic life. Thus, it is not at all difficult to conceive that the two men might have exchanged some of their philosophical ideas.

We have seen that Wittgenstein's use of the term 'phenomenology' and related ideas revolves around the problem of immediate experience and the language that describes it. Also we have examined his use of the term 'primary language' and 'secondary language' and some possible influences from other thinkers. But in order to clarify Wittgenstein's phenomenology to a greater degree, further investigation is needed as to where Wittgenstein obtained the word 'phenomenology'.

2. THE BACKGROUND OF WITTGENSTEIN'S IDEA OF PHENOMENOLOGY

For a clearer picture of Wittgenstein's phenomenology, a consideration of the earlier use of the word 'phenomenology' will be useful. According to Herbert Spiegelberg, the term 'phenomenology' has a rather long history inside and outside the circle of philosophers.[18] On the one hand, Spiegelberg enumerates some ten philosophers who, largely independent of each other, made significant use of the word 'phenomenology', including Kant, Hegel, and Peirce. On the other hand, he lists non-philosophical users of the term, including some surprising names from the field of theoretical physics. What interests us here is the widespread use of the term 'phenomenology' in physical science since the late nineteenth century. Spiegelberg, after mentioning several physicists who used the term, namely, Ernst Mach, Ludwig Boltzmann, Albert Einstein, Max Planck, and Henry Margenau, concludes as follows:

> All these uses are of course related to the merely descriptive, as opposed to the explanatory, conception of physics in G.R. Kirchhoff's school of theoretical physics, for which mechanics was nothing but the attempt to describe the motions occurring in nature as completely and simply as possible without explaining them. Phenomenology in this sense is part and parcel of positivism.[19]

What Spiegelberg says here may very well be right, indeed, except that those physicists he mentions can scarcely be said to have all been positivists in any strong sense of the word. But the fact that there was a standard use of the term 'phenomenology' deserves a closer look with a view to figuring out how the use of the term among physicists might or might not be related to what we generally call phenomenology in philosophy today.

Mach was an especially important representative of the phenomenological approach to physics. As an opponent of atomism in his time, who intended to remove all metaphysical elements in consideration of physics, Mach wanted to establish a universal physical phenomenology that would embrace all domains of physics and facilitate a direct description of as broad a field of physical facts as is possible.[20] Mach believed that our senses are the only source of the immediate revelation of scientific facts and that the world should be described by means of the reduction to sensory experiences without employing any artificial hypotheses. When this approach is applied to the theory of heat, it becomes phenomenological thermodynamics, which operates only with directly observable and thus measurable quantities like pressure, volume, temperature, etc. This method is in sharp contrast to an approach to thermodynamics that goes beyond our sensory experience and employs hypotheses. In statistical thermodynamics, these hypotheses postulate unobservable atoms to explain the phenomena of heat as being statistical properties of large numbers of moving atoms. For Mach, this kind of atomistic approach was not acceptable.

Meanwhile, even though he himself was a proponent of atomistic world-view, Boltzmann discussed sympathetically but critically the phenomenological approach to physics. Boltzmann uses the term in such a way that phenomenology in his sense is only concerned with our experience without any hypotheses in dealing with phenomena. For example:

> [P]hysics must henceforth pursue the sole aim of writing down for each series of phenomena, without any hypotheses, model or mechanical explanation, equations from which the course of the phenomena can be quantitatively determined.[21]

Boltzmann calls this the most extreme form of phenomenology and proposes that this method be mathematical phenomenology because it mainly utilizes differential equations in representing facts. But he also mentions another method called general phenomenology and, not surprisingly, ascribes it to Mach.

> This latter view [general phenomenology] is characterized by Mach's dictum that electricity is nothing but the sum of all experience that we have had in this field and still hope to have. Both views [mathematical and general phenomenology] set themselves the task of representing phenomena without going beyond experience.[22]

Another great physicist, Max Planck, also uses the term 'phenomenology', although in a somewhat different way than Mach or Boltzmann. Planck was once engaged in a polemic with Mach regarding the latter's epistemology. As another proponent of atomism, Planck did not agree with Mach's thesis that the elimination of all metaphysical elements from a physical theory of knowledge is possible. For him, a physicist has to be a realist who searches for what is lasting, unchanging, and independent of everything sensory. Directly opposed to Mach's emphasis on sensory experience, Planck's ideal was to try to uncover what lies beyond sensation.[23]

But this does not mean Planck is entirely opposed to Mach's position, because Planck seems to accept that scientific enterprise begins with experience, even though not all that is real should according to him be reduced to sensory experience. Rather Planck's position is to build on our experience and reach out to what Kant calls 'things-in-themselves' or reality. In a 1941 lecture "The Meaning and Limit of Exact Science,"[24] Planck uses the term 'phenomenological world' in a way that shows it to be identical with the concept of the scientific world picture, by means of which a step-by-step advance of scientific inquiry is achieved. But he also contends that there is a metaphysical sense of the real world that lies behind or even within the phenomenological world, so that a scientific world picture gained by experience—the phenomenological world—always falls short of the metaphysical sense of the real world. Hence, in Planck's use of the word, 'phenomenology' seems to echo the Kantian dichotomy of *phenomena* and *noumena*, one that cannot be overcome.

Finally, Einstein uses the term 'phenomenological physics' to speak of an established method of physics that makes as much use as possible of concepts that are close to experience. He lists names of physicists who employed this approach including Maxwell and Mach.[25] Einstein evaluates phenomenological physics as a tradeoff achieving closeness to experience while losing unity in the picture of the foundations of physics—in contrast to Newtonian physics, which goes beyond the phenomenological view and maintains more unity.

We have thus seen that theoretical physicists often used the terms 'phenomenology' and 'phenomenological physics' to refer to a certain methodological approach to physics. Except for the case of Planck, we can detect a uniform characteristic of these methodological approach denominated phenomenological physics. But more interesting and even striking is the fact that Husserl obtained his term 'phenomenology' from its Machian use. According to Spiegelberg, Husserl was familiar with the Machian use of the word 'phenomenology' as witnessed by his review of Mach's "On the Principle of Comparison in Physics." But what exactly happened between Mach and Husserl seems more crucial than Spiegelberg's simple statement indicates, for as a matter of fact Husserl tells us himself that he was influenced by Mach's use of the word 'phenomenology' and by Mach's quest of a strict scientific method to serve the purpose of a new hypothesis-free science. Husserl admits that phenomenology in the sphere of psychology and philosophy of science was developed as a radicalization of Mach's methodological tendency.[26]

Spiegelberg once called Mach's case pseudo-phenomenology.[27] But that would be right only if Husserlian transcendental phenomenology were the only phenomenology. We can sensibly say that both Mach and Husserl adopted the method of pure description without metaphysical presuppositions or artificial hypotheses. What differentiates Husserl from Mach is that Husserl's phenomenology requires an initial change of attitude, which eventually results in the essential doctrine of the transcendental reduction and, thus, in transcendental phenomenology. But one can also say, more dramatically, that "though Husserl was stimulated by Ernst Mach's attempt, his own phenomenology was formed by the 'transcendental turn' from Mach's mundane phenomenology."[28] Therefore, though it is not generally acknowledged, Mach's influence as a phenomenologist of the natural attitude on the development of Husserl's transcendental phenomenology is critical and, thus, should not be ignored or downplayed, especially not in comparison with the Kantian influence on Husserl which is later and shallower than is sometimes thought.

What has been discussed so far bears an importance with regard to Mach's status as a pioneer of phenomenology, but what interests us at this point is the case of Wittgenstein's phenomenology. It is evident that Husserl admits Mach's influence. Then who, and to what extent, inspired Wittgenstein's phenomenology?

It does not seem that there is any influence of Husserl's on Wittgenstein's phenomenological ideas, in the sense that there is no trace of characteristic Husserlian terminology in Wittgenstein's writings. We have no evidence of Wittgenstein's having ever referred to Husserl or Husserl's works, except for one brief occasion on which Wittgenstein commented on the Husserlian idea of the synthetic *a priori* to some of the members of the Vienna Circle. Prompted by Schlick's question of whether the synthetic *a priori* is possible, Wittgenstein gave a negative answer.[29] Moreover, Wittgenstein did not seem to have a general interest in the transcendental aspect of Husserlian phenomenology. In this sense Wittgenstein's phenomenology seems close to the Machian understanding of phenomenology. But Wittgenstein does not appear to have been directly influenced by Mach in employing his own phenomenology, in that Wittgenstein is reportedly said to have detested Mach's style. Rather more probable is that Wittgenstein became familiar with the term 'phenomenology', or the 'phenomenological method', which is free of hypotheses, from the works or possibly lectures of Ludwig Boltzmann. There is very good evidence for this claim in Wittgenstein's own writings and other documents.

In *Culture and Value* we find that Wittgenstein makes a sort of confession that his ideas are not original but are all influenced by other thinkers. He names some ten thinkers, presumably in chronological order of their influence, starting with Boltzmann.

I don't believe I have ever invented a line of thinking. I have always taken one from someone else. I have simply straightaway seized on it with enthusiasm for my work of clarification. That is how Boltzmann, Hertz, Schopenhauer, Frege, Russell, Kraus, Loos, Weininger, Spengler, Sraffa have influenced me.[30]

According to von Wright, this passage was written in 1931, and it is doubtful that Wittgenstein would have added to the list later in his life.[31] If that is true, and thus if these ten thinkers were so very influential in the shaping of Wittgenstein's philosophy, then Boltzmann's influence must be present in Wittgenstein's writings, even though his name barely appears in them. In fact, Wittgenstein read Boltzmann's works in his teenage years and even intended to study under Boltzmann, only to be frustrated by Boltzmann's suicide in 1906.[32]

If we only have these merely biographical data, we would not yet have reasonable grounds for believing that Wittgenstein was ever influenced by Boltzmann. But, as is shown by the quotes above, Boltzmann's accounts of phenomenology show a striking similarity to Wittgenstein's middle-period characterization of phenomenology. As one of the founders of statistical thermodynamics, Boltzmann could not fully endorse the phenomenological method in physics. It is nevertheless an undeniable fact that Boltzmann uses the term 'phenomenology' in such a way that phenomenology in his sense is only concerned with our experience without any hypotheses in dealing with phenomena. This phenomenological approach to physics, as described by Boltzmann, is closely similar to Wittgenstein's middle-period emphasis on hypothesis-free description of immediate experience. Hence, even though we cannot clearly settle the question of where Wittgenstein got his idea of phenomenology, we can reasonably assume that before he left Vienna Wittgenstein was aware of the use of 'phenomenology' in physics from Boltzmann's works. Of course, this does not mean at all that Mach's ideas are totally absent form the development of Wittgenstein's phenomenological ideas. Wittgenstein occasionally mentions Mach in his writings in a critical manner which shows his familiarity with, and understanding of, Mach's works. For instance, he says that Mach's thought-experiments are at bottom grammatical investigations, which are often identified by Wittgenstein himself with phenomenological investigations.[33]

In view of the fact that Wittgenstein knew Boltzmann's works before he went to England and that his concern was directed toward immediate experience in his early philosophy, it might seem strange that Wittgenstein's first recorded use of the term 'phenomenology' was not until 1929. One possible explanation is in terms of Wittgenstein's general attitude to Boltzmann and Mach. While he admired Boltzmann, Wittgenstein did not value Mach equally highly as Mach's name does not appear in the list of those who influenced him. Boltzmann's position of not giving full support to the phenomenological method in physics might have affected Wittgenstein's avoidance of overt use of the term 'phenomenology' before 1929. But that assumption does not imply that Wittgenstein was not influenced by Boltzmann's ideas of the phenomenological method.

But Boltzmann's criticism of the phenomenological method is not entirely directed at the method itself. He criticizes phenomenologists' uncritical trust in the phenomenological method as a method that contains absolutely no hypotheses. Thus, we could say that his position *vis à vis* the phenomenological method is a rather sympathetic one than one entirely opposed. In a lecture Boltzmann says,

> Phenomenology believed that it could represent nature without in any way going beyond experience, but I think this is an illusion. No equation represents any processes with absolute accuracy, but always idealizes them, emphasizing common features and neglecting what is different and thus going beyond experience.[34]

Boltzmann's point is that even mathematical methods utilizing differential equations for the purpose of describing phenomena will fail to be the absolutely presuppositionless representations which would not violate phenomenologists' dictum never to go beyond experience. No matter how they try to not employ hypotheses or any additional assumptions, phenomenological physicists will end up using some sort of idealization or presupposition that goes beyond experience. Boltzmann gives a specific example in mathematical phenomenology:

> Mathematical phenomenology, as lies in its nature, went back to the idea of continuity of matter, which corresponds to appearances. As against this I have pointed out that the differential equations by definition represent only transitions to the limit, which without supposing the idea of a very large number of individual beings would simply be senseless.[35]

This means that even differential equations employed in mathematical phenomenology would not be possible without an assumption of continuity. So it should rather be understood that Boltzmann did not oppose the idea of phenomenological physics as such. Even though the idea of the atom is admittedly based on hypotheses, Boltzmann apparently is here pointing out that phenomenologists themselves, at the same time they are saying that they do not employ any hypotheses, in fact use an assumption of continuity when they employ the method of differential equations. In fact, there are passages where Boltzmann seems to advance some ideas that could be regarded as prototypes of Wittgenstein's ideas regarding what is given in experience and the way we express it. In his lecture "On Statistical Mechanics," Boltzmann, in discussing the numerous puzzles that arise from the ambiguity of philosophical concepts, says,

> Our task cannot be to summon data to the judgment-throne of our laws of thought but rather to adapt our thoughts, ideas and concepts to what is given. Since we cannot clearly express such complicated conditions except by words, whether written, spoken, or thought in silence, we can say that we must collocate words in such a way as everywhere to lend the most fitting expression to the given, so that the connections we create between words are everywhere as adequate as possible to the connection in reality.[36]

Boltzmann continues that some of the philosophical questions and problems are themselves illusions created by our mental habits, which make it difficult to express adequately the given. Here Boltzmann foreshadows both Wittgenstein's general concern with the relation of language and reality, and, in particular, Wittgenstein's main phenomenological idea of how we describe what is given in immediate experience. Boltzmann's idea of finding the most fitting expression of the given that will create an adequate connection between words and reality comes very close to Wittgenstein's idea of establishing ideal notations that will perfectly describe what is given in immediate experience.

Wittgenstein, who admired Boltzmann, might well have picked up Boltzmann's ideas regarding the ideal of an exact, presuppositionless representation of reality. The problem of representing the phenomenological world, which is immediately given, is what worries Wittgenstein in the *Tractatus*. As far as Wittgenstein's background is concerned, in addition to Russell's theory of acquaintance, he thus takes over the idea of the mathematical and other symbolic representation of reality from Boltzmann. Indeed the whole problem that Wittgenstein struggled with was that of how we can describe what is given in immediate experience. What Wittgenstein eventually claims in the *Tractatus* is that he has shown the possibility of a way of representing the given in terms of a logically ideal language that perfectly describes immediate experience. Hence, the picture idea, which is calculated to show that propositions are pictures of reality and nothing but pictures, is indeed presented as Wittgenstein's solution to the problem of representation. And this solution not only reflects the importance of the problem of representation itself but even more the very phenomenological thesis that the pure description of the given (of the phenomenological world) is possible.[37] When we realize that Wittgenstein's subsequent use of the term 'phenomenology' agrees with Boltzmann's use (i.e., the absolute foregoing of hypotheses), we see that there is no doubt that Wittgenstein was under the influence of Boltzmann's ideas when writing the *Tractatus*.

Apart from the influence from physicists' usage of phenomenology, we could possibly assume some other factors that might have influenced Wittgenstein to use the term. We can consider Rudolf Carnap as another possible influence, though who influenced whom in what respect is not always clear. It is well-known that the Vienna Circle was deeply influenced by the *Tractatus*, which led some of the members of the Circle to arrange meetings with Wittgenstein in the late 20s and early 30s to discuss their philosophical ideas or sometimes to listen to the latest development of Wittgenstein's ideas. Even though the intellectual relationship between Wittgenstein and Carnap did not last long, they knew each other in person, and some sort of intellectual exchange took place between them. An incident in the summer of 1932 revealed that Wittgenstein was highly dissatisfied by Carnap's writings, which he believed contained much of his own ideas without proper acknowledgment. This shows that Wittgenstein and Carnap in fact shared many philosophical interests.[38] Wittgenstein even thought that Carnap had stolen his ideas and published them as if they originated with Carnap or with other members of the Vienna Circle. Even though Wittgenstein's meetings with some of the members of the Circle went on without Carnap after 1929, Waismann, who attended almost every one of these meetings and took notes of them, conveyed Wittgenstein's ideas to the Circle. Therefore, it is not entirely surprising to see Wittgenstein's reaction against Carnap when he saw Carnap's paper that made use of ideas very similar to his own.

What is interesting about this incident is that both Wittgenstein and Carnap indeed shared interests in respective philosophical ideas; either Wittgenstein influenced Carnap, or the other way round. The relevance of discussing Carnap in connection with Wittgenstein's phenomenological ideas is found in Carnap's *Aufbau*, where Carnap, deeply influenced by Russell and by Mach, aims at a construction of the world from the given. While developing his ideas, Carnap not only often compares his ideas with Husserl's phenomenology but uses himself the term 'phenomenological analysis'. It is also interesting to note here that Carnap touches on phenomenological themes, sometimes even using Husserlian vocabulary. More interesting with regard to Wittgenstein's phenomenological ideas, however, is that Carnap's attention is also directed to the immediately given at the beginning stage of his construction program. For example, Carnap maintains,

> At the beginning of the system, the experiences must simply be taken as they occur. We shall not claim reality or nonreality in connection with these experiences; rather these claims will be 'bracketed' (i.e., we shall exercise the phenomenological withholding of judgment', in Husserlian sense.[39]

Even though Carnap's ultimate goal might be different from Husserl's phenomenology, in the sense that Carnap's concern is a rational reconstruction of the concepts of all fields of knowledge, and not in experience or consciousness itself, he consciously relates his analysis of the given to Husserlian phenomenology. In another reference to phenomenology he emphasizes that there are objects which exist independently of being experienced.

> Sensations properly belong to the object domain of psychology; qualities, on the other hand, belong to the domain of *phenomenology* or the theory of objects; there they are called 'objects of sensation'.[40] (Emphasis added.)

This shows that Carnap, by referring to phenomenology, does not merely presents a phenomenalistic thesis, in that what we grasp in our sense-perception (for him 'sensation') is for him different from sense-perception itself and thus not something psychological or mental. The object of sensation is not itself sensation or mere sense-impression, which either belongs to or is dependent on one's mind. To put it differently, it is not a mind-dependent idea, but rather it is genuinely different from any mental act of psychological analysis and cannot be simply confined to their being mere representations of what is real outside our minds.

Hence, what Carnap says about 'the given' not unlike what Wittgenstein basically says of it, is that it is something we grasp when our mental act takes place. Furthermore, Carnap's analysis even shows similarities with Wittgenstein's concern with what is immediately given in one's experience. Carnap says that the given does not have a subject, and he calls his approach methodological solipsism.[41] This point deserves a detailed examination in comparison to Wittgenstein's remarks in his middle-period writings that visual space has essentially no owner. But this is not the place to discuss the problem of solipsism, as I take up this matter later.[42] At any rate, from what we have discussed we can assume that Wittgenstein, as he was in touch with Carnap, may have been reinforced by Carnap's frequent use of the term 'phenomenology' regarding his own phenomenological ideas that had been immanent since his teenage years with the influence from Boltzmann. Indeed, Carnap's philosophical relationship to Wittgenstein can indirectly explain the indirect and unacknowledged Machian character of Wittgenstein's phenomenology, too, because Carnap's *Aufbau* was written under Mach's influence.

3. PHENOMENOLOGY VS. PHENOMENALISM

What has to be noted at this point is that the interests of those who put forward phenomenological method in theoretical physics have nothing to do with an often confused notion of phenomenalism. Mainly from epistemological concerns, a phenomenalist would claim that our knowledge is confined to *phenomena* as opposed to things-in-themselves or *noumena*, for we are directly accessible only to *phenomena*, namely to the sense-impressions of what are supposed to be the real things outside of our minds. But obviously the concern of the above-mentioned physicists is not directed toward the sense-impressions of real things. Phenomenological physics is not the study of sense-impressions. For physicists, the phenomenological method allows them to base their inquiry on the directly observable and thus measurable variables instead of employing any unobservable metaphysical hypotheses such as atoms or molecules. More generally, phenomenology can be characterized as a study that gives primacy to what is immediately given to our experience (i.e., to consciousness) from which the ultimate structure of reality can be revealed. Wittgenstein is doing just this also in his phenomenology, which is not phenomenalistic because it asserts neither that there exists an irreconcilable division between appearance (sense-impressions) and reality (things-in-themselves), nor that we have direct access only to appearance. What we immediately perceive may be mere appearance, and thus we may not be able to penetrate into the true nature of the real thing, but it is also possible that we actually experience what belongs to reality, which is not sense-impressions and is thus totally independent of our minds. Phenomenological investigation is open to that kind of ontological commitment.

Wittgenstein's phenomenology, thus extrapolated from the physicists' idea of phenomenology, is also influenced by the turn of century British realism of G. E. Moore and Bertrand Russell, which also shows a sharp contrast to phenomenalism. In the early phase of realism, Moore, for instance, tried to set up a two-term relation between subject and object in talking about knowledge. It is the relation of the knowing subject and the known object. In "Refutation of Idealism" Moore repeatedly makes a point of differentiating (the act of) consciousness from the object of consciousness. Taking an example of 'blue', Moore goes on to say "'Blue' is as much an object, and as little a mere content, of my experience, when I experience it, as the most exalted and independent real thing of which I am ever aware."[43]

Moore's point is that when we have a sensation, in other words, when we have an act of consciousness, the object of sensation is not a part of our sensation or consciousness but something existing outside our minds. Not only does the object of consciousness exist, but it can exist quite independently of any cognitive acts. By all means, Moore's move is against idealism, which claims that "To be is to be perceived." Moore contends that idealism is wrong simply because those who claim the idealist tenet fail to distinguish sensation and its object. In the case of 'blue', Moore says, idealist do not see the difference between 'the sensation of blue' and 'the existence of blue', and thus they end up claiming that blue does not exist but that only a mind-dependent idea of it does. In other words, Moore's point is that what is experienced is not identical with the experiencing of it. Moore's thesis in "Refutation of Idealism" is not merely limited to an attack on idealism. Asserting that cognitive acts deal with the object of sensation, Moore does not support the phenomenalist idea that we have direct access only to sense-impressions. Cognitive relation is a relation between our minds and what exists outside our minds, i.e., what exists independently of our minds.

A different but similar approach is found in Russell's philosophy in the early 1910s. It is similar in the sense that Russell holds 'what is experienced (or perceived)' is different from 'the experiencing (or perceiving) of it'. But it is different in that 'what is experienced', according to Russell, is sense-datum, whereas 'the experiencing of it' is sensation. What Russell calls sense-data may not be identified with Moore's object of consciousness. But, in character, Russell's sense-data which belong to the major source of knowledge by acquaintance, are not part of our cognitive act. Sense-data are entities independent of the knowing subject. In fact, Russell takes sense-data to be physical entities that should not be confused with Berkeleyan ideas, which are basically mind-dependent.[44]

For Berkeley, when we experience a table, all we have are ideas (i.e., sense-impressions) of table, which are in our minds. But Russell contends, in agreement with Moore, that this is a mistake and the idealist is blamed for failing to see the difference between 'what is experienced' and 'the experiencing of it'. What differentiates Russell's position from any form of idealism, including phenomenalism, is that "the sense-data which we commonly take as signs of the existence of the table are really signs of the existence of *something* independent of us."[45]

A phenomenalist would try to explain our experience and cognition in terms of our minds and sense-impressions, mainly because the phenomenalist believes we are not able to know beyond such sense-impressions. But Russell's point is basically different from phenomenalism, although seemingly similar, for sense-data are not the same thing as sense-impressions. Sense-data are physical, but sense-impressions are mind-dependent. For a phenomenalist, sense-impressions are either all that exists or the representations of things-in-themselves reflected on our minds. On the contrary, Russell's sense-data constitute one source of the most immediate and direct knowledge. Sense-data are what we are directly aware before our minds. Certain patches of color, sounds, taste, smells—which used to be called 'secondary qualities' and were considered not independent of perception—are examples of sense-data. Obviously for Russell, these data are not secondary as opposed to 'primary qualities', let alone mere sense-impressions as opposed to things-in-themselves. Russell contends that the idea that secondary qualities are dependent on perception is due to falsely supposed conception of absolute space as found in the case of Kant. This confirms that Russell does not base his argument on any sort of dichotomy between appearance and reality.[46] The Kantian dichotomy of appearance and reality is certainly not one of Russell's concerns. Instead, Russell attempts to give a firm foundation of our knowledge based on the most direct and immediate data. By saying that sense-data are physical, Russell implies that they are the constituents of the actual world.[47] Hence, we cannot possibly say that Russell's account is one of phenomenalism, which came out as a modified or even a ramified version of idealism.

Wittgenstein as a student of Russell and Moore at Cambridge during their early phase of realism was influenced by their ideas that objects existing independently of our minds are given in our experience. Indeed, Wittgenstein's objects in the *Tractatus* have their origins in Russell's objects of acquaintance to which the above-mentioned sense-data belong as constituents.[48] Influenced by the ideas of phenomenological physics on the one hand, and those of

Moore and Russell on the other, it would make little sense to say that Wittgenstein means phenomenalism when he talks of phenomenology.

This realism of Russell based on objects of acquaintance not only influenced some of the most important ideas of Wittgenstein's *Tractatus* but also shows an interesting contrast to Husserlian phenomenology. One idea that differentiates Husserl's phenomenology from phenomenalism is that his transcendental reduction makes it possible for us to grasp reality directly. Husserl's concern in this regard is to find a way to reconcile our consciousness, which is immanent, with actually exiting things that are transcendent and not inside or within our consciousness. We know that objects transcendent to our consciousness are given in consciousness, which is immanent. But how is this possible at all? This givenness forms one of the important themes of Husserlian phenomenology. So he asks himself again and again in the *Idea of Phenomenology*,

> How can cognition reach beyond itself? How can it reach a being that is not to be found within the confines of consciousness?[49]

> How do I, the cognizing subject, know if I can ever really know, that there exist not only my own mental processes, these acts of cognizing, but also that which they apprehend? How can I ever know that there is anything at all which could be set over against cognition as its objects?[50]

The problem is that in actual cognition, those objects that transcend our consciousness can gain their status as actually existing beings only within the confines of the acts of our consciousness, which is immanent. In other words, we know something is given in our consciousness. To gain the status of the real object, i.e., reality, the things once assumed to be transcendent must become immanent to consciousness, and this is one of the important roles that the transcendental reduction plays in Husserl's phenomenology. According to this scheme, reality directly impinges upon our consciousness, and actual interface between consciousness and reality is set forth as a result.[51] This of course leaves no room for the Kantian dichotomy between *phenomena* and *noumena*, not because Husserl gets rid of the realm of *noumena*, but because what the transcendental reduction amounts to is giving us accessibility to what

Kant calls things-in-themselves. Thus, Husserl could criticize Kant as follows:

> And this is not to say that the things once more exist in themselves and 'send their representatives into consciousness'. This sort of thing cannot occur to us within the sphere of phenomenological reduction.[52]

Thus, Husserl's phenomenology can be free from the possibility of falling into phenomenalism. The given is not only captured in our consciousness, but it is the reality existing in the actual world transcendent to our consciousness. This point is clearly seen in Husserl's own account of phenomenological reduction.

> Rather [phenomenological reduction] entails a limitation to the sphere of things that are *purely self-given*, to the sphere of those things which are not merely spoken about, meant, or perceived, but instead to the sphere of those things that are given in just exactly the sense in which they are thought of, and moreover are self-given in the strictest sense—in such a way that nothing which is meant fails to be given.[53]

By means of the transcendental reduction, we are left with a phenomenological residuum, which is nothing other than consciousness itself that survives the reduction. But also this pure consciousness is left with that which is meant, i.e., the world of the once-assumed transcendent. To put it differently, we are left in the realm of what is immediately given to our consciousness. Husserl makes this point even more clearly in *Cartesian Meditations*.

> This universal depriving of acceptance, this 'inhibiting' or 'putting out of play' of all positions taken toward the already-given Objective world and, in the first place, all existential positions (those concerning being, illusion, possible being, being likely, probable, etc.),—or, as it is also called, this 'phenomenological epoché' and 'parenthesizing' of the Objective world—therefore does not leave us confronting nothing. On the contrary we gain possession of something by it; and what we (or, to speak more precisely, what I, the one who is meditating) acquire by it is my pure living, with all the pure subjective processes making this up, and everything meant in them, *purely as* meant in them: the universe of 'phenomena' in the (particular and also the wider) phenomenological sense.[54]

Of course, here Husserl does not use the word 'phenomena' as opposed to 'noumena' in the way Kant did. Rather, phenomena here is the absolutely and directly given world of data that are presented to the consciousness in the way in which Kantian *phenomena* and *noumena* coincide. Hence, one of the end-result gained through the transcendental reduction includes an actual interface between consciousness and reality that ultimately excludes the possibility of phenomenalism.

Meanwhile, Russell, the realist, also makes a claim very similar to Husserl's with regard to the interface between consciousness and reality. For Russell, sense-data directly impinge upon our minds.[55] Sense-data are not fuzzy images or mere representations of the real world, but are themselves part of the real world. In this sense Russell's project, which amounts to the reduction of scientific knowledge to knowledge by acquaintance and then again the logical construction of the scientific world, leaves room for a very interesting comparison with Husserl's phenomenology. Russell's project indeed parallels Husserl's phenomenological reduction and constitution which secure the ground for an absolute and rigorous science.

CHAPTER II

PHENOMENOLOGY OF THE *TRACTATUS*

1. RUSSELL'S INFLUENCE ON THE EARLY WITTGENSTEIN

In the previous chapter we examined the non-phenomenalistic characteristic of Wittgenstein's phenomenology as influenced by the early phase of Moorean and Russellian realism. This was to make a clear distinction between 'what is experienced' and 'the experiencing of it', which opens the way for a subject to grasp reality directly. That was the position in which Wittgenstein took as a young student at Cambridge. But further examination reveals a deeper philosophical tie between Russell and the young Wittgenstein.

The objects presented in the *Tractatus* are closely related to Russellian objects of acquaintance, and represent what is given in immediate experience. Wittgenstein is concerned with setting forth a pure language that describes what is given in our immediate experience. This attention to immediate experience, perhaps, is one major obsession from which Wittgenstein could never escape throughout his philosophical life. But that kind of problem is possible for Wittgenstein because objects to be described are purely empirical ones. Wittgenstein's faith in the purely empirical might have been even stronger than Russell's, and it eventually resulted in one major disagreement with regard to the status of logical forms. In fact, Wittgenstein could not accept anything akin to Russell's logical forms as separate entities with which to be acquainted.

In general, it is clear that Wittgenstein's objects have their origins in Russellian objects of acquaintance. According to Russell's theory of acquaintance, to be acquainted with something means that something is presented to one's mind or that one is directly aware of something. When we say we know something by acquaintance, we have not laid a claim to truth. We can know something by acquaintance without knowing whether it is true or false. Russell characterizes the nature of (objects of) acquaintance in several places as follows:

> I say that I am *acquainted* with an object when I have a direct cognitive relation to that object, i.e., when I am directly aware of the object itself.[1]

> We shall say that we have *acquaintance* with anything of which we are directly aware, without the intermediary of any process of inference or any knowledge of truths. Thus in the presence of my table I am acquainted with the sense-data that make up the appearance of my table—its colour, shape, hardness, smoothness, etc.; all these are things of which I am immediately conscious when I am seeing and touching my table.[2]

'To be acquainted with' means for Russell that we are 'directly aware of' or 'immediately conscious of' something, and that alone constitutes complete knowledge. In the above example, a particular color we experience, whether it is red or yellow, is itself an object of acquaintance that constitutes perfect and complete knowledge when I see it. No further empirical information about the color is necessary to know the object.[3] In short, Russell's object of acquaintance is what is given in our immediate experience. The way that we pick out this immediately acquainted object is, in the strictest sense, in terms of logically proper names such as 'this' and 'that'. It would be wrong to suppose the proper name 'this' here *means* 'the object to which I am now attending'. Rather, 'this' is a proper name *applied* to the object to which I am now attending.[4] And in this case, the referent, i.e., the object itself to which a proper name is applied, must exist independently to be acquainted and thus to be named. The object of my immediate experience, so to speak, cannot be defined or described but can only be possibly pointed to by naming 'this' or 'that'. As a matter of course, in order to do that, something has to be actually given to me.

These basic ideas of Russellian theory of acquaintance in the early 1910s indeed influenced the young Wittgenstein who became a student of Russell in 1912. A very similar vein of thought is found in one of Wittgenstein's earliest writings. He writes in his *Notebooks 1914-1916*,

What seems to be given us *a priori* is the concept: *This*.—Identical with the concept of the *object*.[5]

This quote might be weak evidence, but clearly, it shows that Wittgenstein follows Russell in the sense that object can only be pointed to by logically proper names like 'this'. Actually, this point is further connected to Wittgenstein's general remarks on names in the *Tractatus*. There, Wittgenstein says a name cannot be defined by further analysis, and the meaning (*Bedeutung*) of it is an object.

> 3.26 A name cannot be dissected any further by means of a definition: it is a primitive sign.
>
> 3.261 Names *cannot* be anatomized by means of definitions.
>
> 3.203 A name means an object. The object is its meaning.

Wittgenstein's observation reminds us of Russell's position quoted above that a name is applied to an object (of acquaintance) that cannot be defined or described. It is not difficult to see that this sort of formulation may well be an outcome of Russell's influence from his distinction between knowledge by acquaintance and knowledge by description. For Wittgenstein, too, objects are the domain of what is given (*TLP*, 2.0124, 4.12721, 5.524), and can only be named (*TLP*, 3.24). Basically, sticking to Russellian ideas of acquaintance, Wittgenstein at the same time deviates from Russell's position with regard to the status of logical form, and this deviation eventually becomes a core of the *Tractatus*.

Russell, in his posthumously published 1913 manuscript entitled *Theory of Knowledge*,[6] tries to put forward logical forms as required elements with which to be acquainted to make a judgment. However, Wittgenstein objected to this idea and criticized his teacher for not realizing that it would not work. The main point regarding this matter is that, according to Russell, to assert a simple relational proposition *aRb*, we not only need to be acquainted with names of particulars like *a* and *b* and that of relation *R*, but also with logical form $x\chi y$, in which each of *a, R, b* is replaced by variables representing the pure form of the proposition *aRb*. Thus we are prevented from judging *bRa* rather than *aRb* when *R* is asymmetrical as in the case of "Socrates precedes Plato." What Russell means by this is that, one way or another, we have to be acquainted with logical forms to understand propositions containing logical concepts, even though he was not sure how this could be done. Russell says in his 1913 manuscript as follows:

For the present, I am content to point out that there certainly is such a thing as 'logical experience', by which I mean that kind of immediate knowledge, other than judgment, which is what enables us to understand logical terms . . . they are understood, and this shows that those who understand them possess something which seems fitly described as 'acquaintance with logical objects'.[7]

I think it may be shown that acquaintance with logical form is involved before explicit thought about logic begins, in fact as soon as we can understand a sentence.[8]

These incredible remarks by Russell indicate that we must already have logical experience, which is not judgment but acquaintance, such that we are able to understand propositions. Furthermore, other than logical forms such as $x\chi y$, which he calls atomic complexes, Russell contends that "there are logical objects which are involved in the formation of non-atomic complexes. Such words as *or, not, all, some,* plainly involve logical notations; and since we can use such words intelligently, we must be acquainted with the logical objects involved."[9] And considering that something has some relation to something else as a logical truth, Russell says the importance of understanding pure form lies in its relation to the self-evidence of logical truth.[10] Rather surprisingly, Russell puts forward the view that acquaintance with logical form cannot involve any falsity.

Russell brings the logical forms of propositions into his discussion in *Theory of Knowledge* because of his theory of acquaintance. According to this theory, any proposition I can understand must consist of objects with which I am directly acquainted. But Russell maintains that acquaintance does not involve knowledge of any truths about its objects, as noted earlier. This idea of acquaintance without any truth claim has a difficulty with regard to the sense and understanding of a proposition. For instance, when we have a proposition *aRb*, we must be acquainted with each object, *a, R,* and *b*, according to this theory, without knowing the truth of each constituent of the proposition. But the problem in this case is that, although we are acquainted with each constituents of the proposition *aRb*, in no way would we also know the sense of the proposition, for mere acquaintance with objects such as *a*, *R*, and *b* will not also give us any knowledge of truth about the objects. And thus, the original theory of acquaintance alone cannot explain the possibilities of objects' combination with each other forming a proposition. In order that we understand the proposition *aRb*, Russell had to bring in an additional premise that makes these objects of acquaintance *a*, *R*, *b* fit together so as to form a proposition. This is the logical form of a proposition, which was represented by $x\chi y$ above.

Basically, this addition of logical form to the objects of acquaintance comes from the very points in which Russell firmly believed. Russell wanted to construct our knowledge from the basis of acquaintance. Acquaintance is the source of immediate knowledge, that is, the ultimate ground on which we build the rest of our knowledge. But, unfortunately, acquaintance with objects alone cannot explain how it is possible that these objects can combine with each other. Russell's solution is to take logical forms as something to be acquainted. This means that the rock bottom of our empirical knowledge represented by objects of acquaintance consists of not only sense-data and universals but also somewhat Platonic entities such as (atomic and molecular) logical forms. His theory reduces the human ability of logical thinking to experience. Without having logical experience first, no logical thought will be possible, according to Russell's scheme. However, these rather Platonic elements within Russell's theory of acquaintance are what Wittgenstein could not accept as being viable, and so he himself presented a revised theory of acquaintance by postulating objects already equipped with logical forms.

When Russell showed a part of his 1913 manuscript to Wittgenstein, Wittgenstein's harsh criticism is believed to have kept Russell's manuscript incomplete and unpublished during his lifetime. The point of Wittgenstein's criticism is decisively shown in one of his letters to Russell dated June 1913 as follows:

I can now express my objection to your theory of judgment exactly: I believe it is obvious that, from the pro[position] "A judges that (say) a is in the Rel[ation] R to b," if correctly analysed, the pro[position] 'aRb. ∨ .~aRb' must follow directly *without the use of any other premises*. This condition is not fulfilled by your theory.[11]

Wittgenstein could not accept such additional premises as logical forms, and as pointed out by David Pears,[12] naturally this point led him to say repeatedly "logic must take care of itself" in *Notebooks 1914-1916* and the *Tractatus*.[13] Also, other evidence shows that Wittgenstein was in fact baffled by the same problem after he had rejected Russell's thesis in the 1913 manuscript. In *Notebooks*, Wittgenstein writes about the difficulty regarding the logical form of a proposition, which unmistakably shows Russell as being the source of the problem:

The fact that in a certain sense the logical form of p must be present even if p is not the case, shews symbolically through the fact that 'p' occurs in '~p'.

This is the difficulty: How can there be such a thing as the form of p if there is no situation [*Sachverhalt*] of this form? And in that case, what does this form really consist in?[14]

As indicated above, Russell's solution to the difficulty, even though he may not have asked the same question, was that the 'form of *p*' is an abstract entity and that one has to be acquainted with it in order to understand *p*. As the passage quoted above shows, Wittgenstein is not satisfied by Russell's solution, and ultimately Wittgenstein presents a different theory based on Russell's theory of acquaintance.

Given that Wittgenstein requires no additional premise, he can then deny the need to be acquainted with logical forms as separate entities. Instead of accepting Russell's idea that in order to understand a proposition we must be acquainted with its logical form, Wittgenstein shows in his early works that when we are acquainted with an object, we must be aware of the possibility of its combining with other objects. When the objects are given, Wittgenstein believes that the possibilities of their logical combination with other objects to form a proposition are also given. Russell's logical form of a proposition, e.g., $x\chi y$, plays the role of providing the possibilities of how objects can combine to form a proposition. This is the form that lets us know whether the objects we experience belong to the right type, and therefore are able to fit together. On the other hand, for Wittgenstein, this additional entity is unnecessary. Wittgenstein clearly views that, if we are acquainted with objects, we also come to know what possibilities these objects have in order to form logical relations with other objects. The ways in which objects are combined together no longer depend on the logical forms of propositions. But somehow these forms are built into the objects. The ways in which objects are combined are thus written into the objects themselves. Ultimately, Wittgenstein presents objects with built-in structure of logical forms in the *Tractatus*.

> 2.012 In logic nothing is accidental: if a thing *can* occur in a state of affairs, the possibility of the state of affairs must be written into the thing itself.
>
> 2.0123 If I know an object I also know all its possible occurrences in states of affairs.
> (Every one of these possibilities must be part of the nature of the object.)
>
> 2.014 Objects contain the possibility of all situations.
>
> 2.0141 The possibility of its occurring in states of affairs is the form of an object.

Therefore, Wittgenstein's logical form of an object is a form that determines the possibilities of how its object is combined with other objects to form a state of affairs and, thus, to make sense of a proposition. Unlike Russell, Wittgenstein believes that we do not have to be acquainted with logical forms. All we have to do is to be acquainted with objects in which logical forms are inherent. For Wittgenstein, the ability of logical thought is not based on logical experience. Because of the convenient feature of objects, which are already equipped with logical forms, all we need in order to understand propositions is to be acquainted with objects. Indeed, this is a strange variation of Russell's ideas. Wittgenstein's position is that we need not be acquainted with logical forms. Yet, once objects are given, logical forms are also given at the same time. By means of experiencing objects, we automatically experience logical forms too. But certainly it is not experience in the sense of what we can call 'logical experience' as Russell called it. All that is needed in order to understand logic is to experience objects, and not logic. In other words, objects must be given in immediate experience. According to Hintikka and Hintikka, this is the exact point that Wittgenstein addresses in the *Tractatus* 5.552 against Russell.[15] Indeed, we do not need any experience in logic after eliminating the Platonic elements with which to be acquainted.

Many other detailed answers to Russell are found here and there. For example, Wittgenstein says in *Notebooks* that "the logical form of the proposition must already be given by the forms of its component parts."[16] Since the component parts of a proposition are names (*TLP*, 4.22) and names mean objects (*TLP*, 3.203), the preceding remark can be paraphrased as "the logical form of the proposition must already be given by the forms of its objects." Truly, this remark is another way of putting forward Wittgenstein's view that the logical forms of objects determine the possible ways in which objects can be linked together to form a state of affairs. Indeed Wittgenstein modifies Russell's theory of acquaintance by eliminating logical forms as separate entities with which to be acquainted and to put them into objects. The consequence of this modification is enormous in the subsequent development of Wittgenstein's ideas. A few days after the previous remark was written, Wittgenstein puts down the followings:

> The knowledge of the representing relation *must* be founded only on the knowledge of the component parts of the situation [*Sachverhalt*]![17]

> One name is representative of one thing, another of another thing, and they themselves are connected; in this way—like a *tableau vivant*— the whole images the situation.[18]

Wittgenstein now seems to hold more than a rough idea of a corresponding relation between name and object. In order that we are able to say something is a constituent of a proposition, Wittgenstein would ask whether it has a corresponding counterpart in states of affairs. Now, according to Wittgenstein, logical forms are explained away as inherent in simple objects, and thus we do not have to bother looking for their counterpart. Yet, there are other type of logical objects which Russell wants to take as objects of acquaintance. They are the so-called logical constants that enable us to make sense of and understand complex or molecular propositions. Even though

Wittgenstein reaches the idea of a corresponding relation between name and object in *Notebooks*, he does not explicitly deny logical constants as non-representatives. Rather, Wittgenstein seems to struggle with the status of logical constants.

Roughly speaking: before any proposition can make sense at all the *logical* constants must have reference.[19]

Unlike the logical forms of atomic complexes about which he could state his position clearly in his letter to Russell as early as June 1913, Wittgenstein still battles with the status of the logical forms of molecular complexes, i.e., logical constants, as late as October 1914. Yet, in the *Tractatus* where he develops a full-fledged thesis of the logical forms of objects as sole logical forms, Wittgenstein rejects Russellian logical objects of molecular complexes.

> 4.0312 My fundamental idea is that the 'logical constants' are not representatives; that there can be no representatives of the *logic* of facts.
>
> 4.441 There are no 'logical objects'.
>
> 5.4 At this point it becomes manifest that there are no 'logical objects' or 'logical constants' (in Frege's and Russell's sense).

This observation clearly means that, only with the logical forms of simple objects, we must and are able to perform all the jobs that Russell's separate logical objects have done. For Russell, logical constants, too, were logical objects with which to be acquainted. But Wittgenstein, eliminating them as non-representatives, tries to reduce all the complex logical forms to the logical forms of simple objects. If the role of logical constants is to make sense and in order to understand logical complexity such as complex or molecular propositions, the denial of logical constants is in a way a natural consequence of the emphasis on objects with a built-in structure of logical forms. According to this scheme, all logic is reduced to the logical forms of objects. Indeed, this incredible consequence of not basing logic on Russell's unusual sense of 'logical experience' is discussed by Hintikka and Hintikka as follows.

Wittgenstein thinks that the complex logical forms are all determined by the logical forms of elementary (atomic) propositions. Furthermore, the logical forms of atomic propositions are determined by the forms of objects. It is via this two-stage process that all logical forms can be reduced to the logical forms of simple objects. From them, all possible logical forms can be constructed.[20]

However, what is important to emphasize in Wittgenstein's thinking here is the idea that something independent of my mind must be given initially in immediate experience not only as the building-blocks of the world but at the same time as a source of the possibility of logical thinking. In other words, the actual givenness of objects, which bears the possibility of logical atoms, has to be initiated in order that my language has meaning at all. And the ground for this givenness is already found in Russell's account of acquaintance:

> 'This' is not waiting to be defined by the property of being given, but is given; first it is actually given, and then reflection shows that it is 'that which is given'.[21]

What all this discussion shows is how closely Wittgenstein's early philosophical ideas are related to Russell's philosophy of the early 1910s, and also how Wittgenstein departs from Russell's position while planting some of the most important ideas of the *Tractatus* in his soil. The full understanding of this philosophical relation between the two thinkers and the direction that Wittgenstein takes will prove crucial in understanding Wittgenstein's phenomenology implicit in the *Tractatus*.

2. THE NATURE OF TRACTARIAN OBJECTS

An obstacle to the interpretation of Tractarian objects as what is given in immediate experience is that Wittgenstein does not explicitly says this in the *Tractatus*. In fact, he barely spells out what objects really are. He expresses what he has to say about objects in an oblique way mainly because of the ineffable nature of objects. For Wittgenstein, objects are the domain of the given, as in the case of Russellian objects of acquaintance, and thus they can only possibly be named but not described or defined. The only way to pick them out is by pointing to them and saying 'this' or 'that'. We can say how objects are combined together to form the world, because all the possible logical forms that determine how objects are linked together to form the world are also given with objects. Wittgenstein shows this in some of the propositions in the *Tractatus*.

> 1 The world is all that is the case.
>
> 2 What is the case—a fact—is the existence of states of affairs.
>
> 2.01 A state of affairs (a state of things) is a combination of objects (things).
>
> 2.0124 If all objects are given, then at the same time all *possible* states of affairs are also given.

Indeed, the world consists of all the possible states of affairs, namely all the possible combination of objects. By figuring out how the logical forms of objects operate in language, therefore, we can possibly say how the world is arranged but neither what the world is nor what those objects are. Logic is a matter of *a priori*. Insofar as this a priori nature of logic is concerned, we do not need experience at all. But the question of what objects are is an empirical one that cannot possibly be answered in an *a priori* manner. That objects have to be given in our experience is what Wittgenstein eloquently expresses in the *Tractatus* 5.552:

> 5.552 Logic is *prior* to every experience—that something *is so*.
> It is prior to the question "How?", not prior to the question "What?"

Hence, we are not supposed to ask or answer the question of "What is an object?", for this question concerns a totally different matter than what logic deals with. It concerns empirical matter and can only be *shown* by pointing to actual experience. Being aware of this point, Wittgenstein could neither give an explicit explanation of the nature of objects nor examples of them. Yet, as time passed by what Wittgenstein purports to say about Tractarian objects is indirectly shown by his own criticism of his earlier view and also by the elaborated words recorded by his close philosopher friends such as Friedrich Waismann.

In the early thirties, Waismann wrote "Theses" in an attempt to interpret some of the main principles of the *Tractatus* and to present them in a comprehensive form. As mentioned earlier, Waismann was one of the members of the Vienna Circle who had meetings with Wittgenstein and was later credited by Wittgenstein as performing a role in conveying his ideas to the Circle. Furthermore, Waismann was working on a book, the announced title of which was *Logik, Sprach, Philosophie*, a would-be introduction to the *Tractatus* upon Wittgenstein's approval. According to the Editor's Preface to the *Ludwig Wittgenstein and the Vienna Circle,* where Waismann's "Theses" is included as Appendix B, the "Theses" was at one time intended to form a part of *Logik, Sprach, Philosophie.*[22] Hence we may say that Waismann's "Theses" contains a great deal of explicative material about the contents of the *Tractatus.* In fact, some of the remarks in the "Theses" directly support what I have argued regarding the status of Tractarian objects. Waismann writes,

> The signs occurring in elementary propositions are called *primitive signs* (elementary signs).
>
> The primitive signs cannot be analysed by definitions.
>
> The meaning of primitive signs can only be indicated by pointing.[23]

In the above quote, there is no doubt that primitive signs in elementary propositions are names in the *Tractatus* that pick out objects. As Wittgenstein himself says in 3.26 and 4.22 of the *Tractatus*, a name is a primitive sign that cannot be dissected or anatomized by definition; its immediate combination makes up an elementary proposition. This in fact proves that Waismann's "Theses" is aimed at the comprehensive presentation of the main principles of the *Tractatus*. In this context, the last sentence of the above quote adds an important insight in understanding Wittgenstein's aphorism-like presentation, which misled many, because in addition to paraphrasing 3.26 and 4.22, Waismann's account assures the nature of these names: the meaning of names cannot be described or otherwise explained because names are simple signs; the meaning of names can only be indicated by pointing. The very Russellian point not clearly stated in the *Tractatus* is revealed through Waismann's testimony.

As in the case of Russell's logically proper names, what the names of the *Tractatus* stand for can only be shown by pointing to an object. This one important clue reveals that the relation between name and

object is indeed a pointing relation reflecting a direct influence from Russell's theory of acquaintance.

This naming relation as a constituent of elementary propositions is further elaborated by the point that what is described by names and propositions is immediate experience. Waismann's "Theses" says:

Phenomena (experience) are what elementary propositions describe.[24]

Elementary propositions describe the content of our experience. All other propositions are merely an expansion of that content.[25]

In other words, Waismann's "Theses" shows that elementary propositions describe what is given in immediate experience, the very phenomenological point of the *Tractatus*. As names are constituents of elementary propositions and names pick out objects equipped with logical forms, which determine the whole logic of propositions, to say elementary propositions describe the content of experience is the same thing as stating that names pick out the objects of experience. Indeed, this idea reveals the phenomenological characteristic implicit in the *Tractatus*. It is the logical forms of objects that determine the way in which the world is arranged, i.e., the entire logic of how the world is. Thus, the possibility of understanding the entire logical structure of the world can be found in immediate experience.

Another evidence in Wittgenstein's own words supports our idea that the naming relation is a pointing relation. In the *Brown Book*, with the newly developed ideas of language-games, Wittgenstein criticizes one of his earlier mistakes saying "the danger of delusion which we are in becomes most clear if we propose to ourselves to give the aspects 'this' and 'that' names, say A and B."[26]

Of course, this idea would be further developed as ostensive definition playing only a part within the system of language-games. However, at the same time, we can directly see what went wrong with his ideas in the *Tractatus*: namely, he once thought that 'this' and 'that' were names in a pointing relation with objects.

Only with this handful of evidence does it seem obvious that the *Tractatus* contains phenomenological themes. With a revised theory of Russellian objects of acquaintance, Wittgenstein could not help developing his phenomenological idea that our entire conceptual structure depends somehow on our experience, an idea that actually enables us to make sense of rather obscure propositions that sound cryptic otherwise. Wittgenstein says,

> 5.6 *The limits of my language* mean the limits of my world.

This 'world', when understood as the one composed of the whole combination of the objects of my experience, will necessarily have its limits in the limits of my language. As was noted earlier, our language cannot start out as making sense unless the objects of experience as the atomic unit of meaning are initially given, precisely because those objects are equipped with logical forms necessary to understand propositions and this language. Hence we can say that where lie the limits of my language, there lie the limits of my world. And 5.61, which explains 5.6, only elaborates the same point:

> 5.61 Logic pervades the world: the limits of the world are also its limits.

This aphoristic assertion would make little sense if we interpret the nature of objects and logic otherwise than what has been argued. But once we see the nature of objects equipped with logical forms as determining the whole logic of language and thus the way the world is, then the proposition 5.61 is a very natural consequence to assert. The world is composed of the objects of immediate experience. But as the logical forms of these objects determine the entire way in which the objects are linked together to form the world, it is no wonder at all to say logic pervades the world. Consequently, logic and the world share limits. One of the most remarkable consequences of the Tractarian scheme, however, is expressed a few lines later in propositions 5.621 and 5.63.

> 5.621 The world and life are one.

> 5.63 I am my world.

If objects are not given at all, or are just mind-dependent in nature, then the above remarks will make absolutely no sense. Something independent of our minds must be given in experience in order that we build our conceptual system from those objects to the entire world. The ultimate building-blocks are required, and those are found in my immediate experience. By means of their being given with logical forms, which constitute the logical atoms of meaning in my language, I can make sense of my language and the world I live in, i.e., my world. This identification of 'I' and 'my world' reveals Wittgenstein's whole point that the objects are the content of my experience.

Now, if Tractarian objects are the objects of immediate experience, which are given to us, and the logical forms of these objects determine all other logical forms and thus not only the way that objects are combined together but the way that the world is arranged, then the logical forms of the objects ultimately constitute the whole logical structure of the world. Being the building-blocks of the world, objects contribute to the way that the entire logic of our language is formed. The possibility of understanding the entire logical structure of the world, then, solely lies in our immediate experience.

This important point that our conceptual structure can be revealed by virtue of paying attention to immediate experience is later made clear by Wittgenstein's own account in one of his lectures. Wittgenstein is reported to have said,

> Sense-data are the source of our concepts: they are not caused by our concepts.[27]

> The world we live in is the world of sense-data.[28]

In Russell's theory of acquaintance, sense-data are the crucial components of the objects of acquaintance. Even though we need more qualifications on how Wittgenstein uses the word 'sense-data', we can reasonably acknowledge, from the above quotes, sense-data as what belong to the realm of objects with which to be acquainted. Then, Wittgenstein's above remarks simply support his phenomenological viewpoint of the *Tractatus*, where simple objects are given in experience to provide the logical atoms of linguistic meaning, the building-blocks of the world, and thus the source of our conceptual structure. Our concepts do not cause sense-data. On the contrary, it is sense-data that cause our concepts. Thus, the sense-data given us play their role as the source of our concepts. This all comes down to saying that everything meaningful in our conceptual scheme comes from our immediate experience, and this is the heart of the phenomenological viewpoint.

As Russell was not interested in or influenced by the Kantian dichotomy of *phenomena* and *noumena*, it is highly unlikely that Wittgenstein, whose view on objects is a variation of Russell's object of acquaintance, was putting forward any theory that reflects the Kantian distinction of things-in-themselves and their appearance. Nowhere in Wittgenstein's early and middle-period writings do we find remarks showing that Wittgenstein was more interested in the given phenomena as opposed to things-in-themselves, which are considered unknowable in nature. Wittgenstein's use of sense-data, as it is initially influenced by Russell, does not allow the interpretation that his concern is rather confined to phenomena than the world of things-in-themselves that is left unknowable. How the early Wittgenstein uses the term 'sense-data' is indirectly shown by his colleague and close friend, Frank Ramsey's hand-written records of conversations:

> W says [it is] nonsense to believe in anything not given in experience . . . .For to be mine, to be given in experience, is [the] formal property to be a genuine entity. Other people's s.d.[=sense-data] and my s.d. i.e., those attached to my body as other people's to their bodies are logical constructions. . . .Take the criterion of what he sees to the meaning of [what] he sees. Similar I see—I see. Then I see is also connected with reality (visual appearance) [in a way] other than how his seeing is, and this is [the reason] that the world is my world.[29]

First of all, this quote decisively proves that Wittgenstein really viewed what we have in immediate experience to be 'genuine entity', with sense-data not being the representation of anything real. To be a genuine entity, it has to be given in immediate experience. Wittgenstein does not first presuppose things-in-themselves and their appearance before he says appearance is real. Rather, he sees that our sense-experience gives us the way we grasp reality. Consequently, we can say that Wittgenstein has no philosophical interest in the distinction between the noumenal and the phenomenal. Wittgenstein's account might be mistaken for phenomenalism, but the background of his philosophical problems is not that of Kant but is closely tied with that of Russell. Also, from the above-quoted record by Ramsey we can see exactly what Wittgenstein means by the statement "the world is my world" in 5.62. It unmistakably shows that the totality of experience makes up the world. The totality of experience here consists of the objects of immediate experience, which are nothing other than sense-data as Ramsey reports.

What has been discussed decisively shows that the entire scheme of the *Tractatus* is nothing other than phenomenological. The objects are phenomenological objects, and the ideal language postulated with logically correct notations is phenomenological language that faithfully describes what is given in immediate experience. Indeed, what concerned Wittgenstein most was finding the proper notations that mirror the phenomenological world.

3. HUSSERL, RUSSELL, AND WITTGENSTEIN

In the preceding sections, I examined how strong the philosophical bond between Russell and Wittgenstein really is in terms of the theory of acquaintance, which is worked out as Wittgenstein's own idea of phenomenology. As briefly argued in the last part of chapter one above, the realist position of Russell can be interestingly compared with Husserl's phenomenology. Husserl's phenomenology not only developed out of the radicalization of the Machian idea of phenomenology, but it also shows a similarity with Russell's realism in addressing what is directly and immediately given. As we have seen, Russell's influence on Wittgenstein's *Tractatus* is obvious, and further comparison between Husserl and Russell might warrant a better understanding of Wittgenstein's phenomenology.

As was argued earlier, what we gain by the transcendental reduction is an actual interface between our consciousness and reality. By bracketing out that which is transcendent, not only are we left with the most direct and immediate givenness, i.e., our pure consciousness, but also with the possibility of having those supposed transcendent objects as actually existing reality within the confines of our consciousness. What Husserl proposes is the annihilation of the world by performing the phenomenological reduction. But that does not mean we lose anything by that. We rather gain the 'constituted' world, which is nothing but reality overlapped and thus found in the realm of pure consciousness by means of the phenomenological epoché. In what the transcendental reduction results is described by Husserl as follows:

> [W]e put all those positings 'out of action', we do not 'participate in them'; we direct our seizing and theoretically inquiring regards to *pure consciousness in its own absolute being*. That, then, is what is left as the sought-for '*phenomenological residuum*', though we have 'excluded' the whole world with all physical things, living beings, and humans, ourselves included. Strictly speaking, we have not lost anything but rather have gained the whole of absolute being which, rightly understood, contains within itself, 'constitutes' within itself, all worldly transcendencies.[30]

This is one very interesting feature of Husserlian phenomenology. A subject performs the transcendental reduction, and everything is put out of action. All the judgment about that which has existed thus far is suspended. First and foremost, what survives this epoché is the pure consciousness of the subject, which in fact, itself performed the epoché. Thus, this phenomenological subject, i.e., pure consciousness, does not belong to the usual sense of the world that has been overthrown. However, this pure consciousness is not the sole phenomenological residuum that survives the transcendental reduction. This pure consciousness is at the same time a constituting consciousness in which the noetic-noematic correlation and thus the possibility of the world of noemata is found. Consciousness, therefore, not only acts as a performer of putting the transcendent world out of action but also as a locus where a new sense-bestowal activity of noesis takes place to constitute the meant world. Since it does not belong to the world, pure consciousness is totally independent of the world. It is not a part of the world. But at the same time, the actual interface between this consciousness and the real world is gained through the course of the transcendental reduction. In a constituting consciousness the legitimacy of the meant world is secured.

In short, the phenomenological reduction provides us with an ultimate basis for the constitution of the meant world, i.e., reality. The pure consciousness resulting from the phenomenological reduction is constituting consciousness, in which continuous acts of intention are found, and thus the world with a whole new meaning is brought about.

In that case, what do we find in this realm of pure consciousness? What really constitutes reality that impinges upon our pure consciousness? To put it differently, what is the nature of Husserl's phenomenological objects? What is immediately given as being captured in pure consciousness is what Husserl calls 'hyletic data', which are the formless raw materials that constitute one component of pure intentional consciousness. Sometimes, Husserl also uses the terms 'hyle' or 'sense-data' for this outcome of the transcendental reduction. However, the phenomenological reduction does not give us these hyletic stuff alone. Along with these unstructured raw materials, we are also given the noetic forms. Husserl explains this as follows:

> The characterization of the phenomenological reduction and, likewise, of the pure sphere of mental processes as 'transcendental' rests precisely on the fact that we discover in this reduction an absolute sphere of stuffs and noetic forms whose determinately structured combinations possess, *according to immanental eidetic necessity*, the marvelous consciousness of something determinate and determinable, given thus and so, which is something over against consciousness itself, something fundamentally other, non-really inherent, transcendent.[31]

The absolute sphere of stuff can be structured as objects by noetic forms that are characterized as our intentional mental activities. So the unformed hyletic data cannot become phenomenologically proper objects before the constituting noetic activities take place. They are not categorialized on their own but are in need of 'sense-bestowing' activities by noesis. Indeed, one important feature of Husserlian intentional phenomenology is that consciousness is always consciousness of something. Our noetic activity performs exactly the intentional function of pure consciousness. It transforms a raw material given in experience into noema, which is inconceivable without its correlation to noesis. In other words, what noetic activity does is to make unarticulated experience into an experience of something, i.e., as something intended. This noetic activity is made possible only by imposing its essence on the other component of pure consciousness, hyletic data.

In fact, this is one reason that Husserl needed such a thing as categorial intuition (and later eidetic reduction), which immediately furnishes a noesis with forms or essences, in terms of which a noetic activity works on hyletic data to constitute a phenomenological object. However, these formless data as such cannot be captured separately, because what can be cognitively meaningful has to be something intended in our consciousness. In other words, without noetic activities being imposed upon, data given by sensory perception cannot be phenomenologically proper objects. Without having constituted activity first, we cannot even refer to hyletic data as 'hyletic data', because before the intentional meaning is bestowed, data alone cannot yet be sensibly discussed. Thus, at first Husserl needed categorial intuition (or eidetic reduction) to constitute reality. And this idea of constitution is what makes Husserlian phenomenology prone to idealism. After the transcendental reduction, the step-by-step reductions of other sorts must follow to carry out constitution. What Husserl finally has as a phenomenological object, therefore, is not an empirical object that he zero-indexed all along with the transcendental reduction. There can be no phenomenological objects in the strict sense of empirical objects for Husserl. Phenomenological objects for Husserl are constituted ones in terms of the noetic-noematic correlation furnished by categorial intuition.

On the contrary, Russell's case is better articulated and clearer with regard to what is given to our minds. For Russell, objects of acquaintance are the immediately given in our experience. These objects include not only sense-data but also universals and even logical forms as presented in his 1913 manuscript *Theory of Knowledge*. So if Russellian reduction is a reduction to acquaintance, which as in Husserl's case is the securing point where the actual interface between mind and reality takes place, then this reduction is categorially articulated, a differentiating feature of Russell's position from Husserl's case. Hintikka points this out as follows:

In Russell, the end-points of reduction to acquaintance are entities of a well-defined categorial status, and in some cases even having an internal structure. They include particulars (individuals) like sense-data, but they can belong to other well-defined logical types, such as different types of universals, logical forms, etc.[32]

This means that what is given in Russell's case is, unlike what we saw in Husserlian reductions, given in the manner that it is already categorially structured. It is either particular like sense-datum or the members of other logical types like universals or logical forms with which we are acquainted.

Indeed, this is one non-trivial difference between Husserl and Russell, mainly because this is the key feature that makes Russell the realist as a realist. Russell does not have to emphasize any activity on the subjective side in explaining our conceptual structure. Sense-data impinge upon our minds, and nothing much is needed to cognize them as particular data such as yellow or blue. On the other hand, for Husserl noetic-noematic correlation, in which form-giving activities on hyletic data take place, is necessarily required to articulate object as object. No naively given objects are phenomenologically meaningful for Husserl. An empirical object transcendent to consciousness has new meaning as phenomenological object only after some articulating work on the subject side are imposed upon. Inevitably, this Husserlian scheme is prone to idealism, for we cannot grasp phenomenological objects with sensory intuition or hyletic data alone.

If we push this line of respective difference between Husserl and Russell, something more interesting can be revealed. Since Wittgenstein's Tractarian objects are closely related to Russellian objects of acquaintance, and represent what is given in immediate experience, an interesting contrast between Husserl and Wittgenstein can be set forth. In his phenomenology, Wittgenstein's concern is the language that describes what is given in our immediate experience. Indeed, what Wittgenstein attempts to do in the *Tractatus* is to find suitable logical notations, i.e., logical analysis, which correctly describe the world of experience. But that kind of a problem is possible for Wittgenstein, because phenomenological objects to be described are purely empirical ones. What Wittgenstein addresses is not like Husserl's constituted reality, which contains idealistic elements. As I discussed earlier, Wittgenstein could not even accept

Russell's idea of logical forms as separate entities with which to be acquainted. In fact, Russell's logical form is a sort of form or essence, an abstract entity that makes logical sentences meaningful. Even though Russell spells out that logical objects are something with which to be acquainted, he himself could not throw light on how this might be possible. On the contrary, Wittgenstein, realizing the unviability of Russell's solution since Wittgenstein himself had also tried out that option, puts forward another unique solution, i.e., putting logical forms into the objects, an idea which ultimately resulted in Wittgenstein's phenomenology.

However, what is interesting, phenomenologically speaking, is that this disagreement in the status of logical forms puts Wittgenstein on one side and Russell along with Husserl on the other, as Russell's 1913 inclusion of logical forms as objects of acquaintance can be viewed as a counterpart to Husserl's categorial intuition.[33] Because categorial intuition can be paraphrased as immediate 'seeing' of universal essences that enables us to have categorialized cognition, it could be thought as an equivalent to Russell's logical forms that identify *aRb* as *aRb*. Even though Russell does not clearly explain how this is possible at all, the position that we are somehow acquainted with logical forms or logical objects amounts to saying that logic should be intuited or that we must have something similar to logical intuition in order to make sense of propositions containing logical concepts.[34]

Hence, the fact that Wittgenstein does not have and does not need logical forms (of Russell) or categorial intuition (of Husserl) naturally makes Wittgenstein's phenomenology a different kind of phenomenology from Husserl's. Unlike Husserl, what Wittgenstein needs is not phenomenological reduction but a language that picks out already articulated phenomenological objects in our immediate experience. As a revision of Russellian theory of acquaintance and a variation of Russellian reduction to objects of acquaintance, Wittgenstein may have reduction to simple objects as the building-blocks of the world. But more weight has to be given to the language that is supposed to describe those objects and thus the world. As a result, what Wittgenstein carries out in the *Tractatus* is logical analysis of an ideal language that consists of propositions in which names play a role in picking-out those simple objects. Hintikka puts this idea as follows:

> [I]n Wittgenstein there is no unformed data and no free-floating logical forms to bring to bear on something. Therefore, all that a reduction can do, if we think of the process on the level of sentences, is to take us from complex sentences to the elementary ones, and from elementary sentences to their ingredients, the simple names. (Since language and thought operate in tandem in the *Tractatus*, this logical reduction reflects the conceptual reduction which is at issue here.) This is precisely what Wittgenstein seeks to establish through his pictorial analysis of elementary propositions and through the extension of the picture theory to all other proposition through the truth-function theory. These are hence Wittgensteinian counterparts to phenomenological reductions, *mutatis mutandis.*[35]

In Wittgenstein's case, unlike in Husserl's, there is no unstructured raw material on which noetic activities gives the forms. But also unlike Russell, there are no logical forms for the requirement of logical thinking. For Wittgenstein, as far as logical form or essence is concerned, there is nothing out there waiting to be acquainted or intuited. All that is required for phenomenological research in Wittgenstein's case is to be acquainted with simple objects, which are already equipped with logical forms and are already classified into logical types. Wittgenstein's attempt is to base logic in its entirety on the logical forms of simple objects, whose existence is necessarily postulated to play the role of logical atoms of our conceptual world. For Wittgenstein, in order to gain an initial meaning of our language, objects have to be given in our immediate experience. The ultimate basis of the entire conceptual structure is found in the experience of being given. Hence, in Wittgenstein this relation to the simplest possible ingredient of the world and language, namely object and its name, is indeed comparable to Husserlian reduction as far as the possibility of meaning is found in the givenness of objects well equipped with logical forms.

This is why Wittgenstein emphasizes phenomenological language rather than a Husserlian apparatus such as phenomenological reduction. This also indirectly explains why Husserl does not seriously develop ideas or theory of phenomenological language. It is because, for Husserl, phenomenologically proper objects are constituted ones. What is given before noetic activities take place is not an object but formless raw material that cannot be picked out or named like the Tractarian object. Husserl cannot have phenomenological language that describes what is given in pure experience because for him, what is given in pure experience is hyletic data, which cannot be picked out in any language. Since there cannot be any phenomenological objects in an empirical sense, Husserl cannot have a purely phenomenological language. And this explains Husserl's inevitable concession to idealism.

4. PHENOMENOLOGY AND SOLIPSISM

As I have illustrated, phenomenology can be characterized as a study that amounts to the reduction of all that is cognitively meaningful to immediate experience. Basically, phenomenology is a study that gives primacy to what is immediately given to our experience or consciousness. It maintains that the basis for a meaningful discourse of genuine knowledge should be found in immediate experience. The idea that we have to attend to immediate experience to have genuine knowledge shows that a sort of solipsism is built into the idea of phenomenology. However, the solipsism implicit in phenomenology is not a kind of solipsism that doubts the real existence of other subjects or other minds. It is rather a methodological solipsism that cannot be avoided in the phenomenological position.

One of phenomenology's most important features is the attention to our own immediate experience. To delve into immediate experience, what we should deal with necessarily is the ego, which plays the role of a subject in which all experiences are experienced. The ego serves as the pole of all experiences. It is the subject of immediate experience. It can also be said to be the mind or consciousness, which is considered to be distinguished from the object of consciousness. This distinction between subject and object seems inevitable when we hold a purely empirical position, for there has to be some place from where all experiences are viewed as experience. A significant characterization of this subject of experience is found in Moore's "Refutation of Idealism."

According to Moore, the subject (i.e., mind or consciousness) is a place where the mental act takes place. By means of this act of consciousness, the subject obtains the object of consciousness. What is important in this regard is that the act of consciousness and the object of consciousness are distinct. Something exists outside of one's mind (i.e., one's consciousness). The role of this subject part is to apprehend what is given from outside it. When one sees the color green, green itself is something outside one's consciousness. It is known to consciousness through the sensation of green that one has in one's consciousness. It is only when there is a sensation of green that the subject plays a role and the object is known, so that one becomes conscious of green as green in one's mind. Moore's distinction between the act of consciousness and the object of consciousness is revealing with regard to the phenomenological standpoint that emphasizes one's attention to immediate experience. As long as attention is directed toward what is given in one's immediate experience, one's ego should be treated as detached from that which is experienced, because it is the point to which every experience is given. In other words, to employ Moore's terminology, the ego is not the object of consciousness, but is itself consciousness in which the mental act takes place to have the object of experience. However, the built-in difficulty of this position is that we cannot avoid solipsism insofar as the ego serves as the pole of all meaningful experience. Because the ego is the place where the objects of experience are given, the ego itself cannot belong to the objects of experience. And also, because these objects themselves exist outside the ego and become the objects of experience through the mental act of the ego, these objects lose their meaning without the ego's playing a part as the owner of experience.

Here is where the problem of solipsism comes into play. In this position the objects of experience have to be the objects of the ego which is the owner of the experience. In other words, when I experience objects, those objects have to be my objects. Solipsism happens when a great deal of emphasis is given to immediate experience, in which case all one can accept is one's own experience and not others', for another ego or its experience cannot be directly known to me. Hence, we may say that solipsism is implicit in the phenomenological position that gives primacy to one's immediate experience. This methodological solipsism is common to Husserl, Russell, and Wittgenstein. I discuss how this problem arises in the general scheme of phenomenology in each case as follows.

In Husserl, what the transcendental reduction is designed to achieve is to have us place ourselves in the realm of the immediately given to our consciousness. By means of the transcendental reduction, we are left with a phenomenological residuum, which is nothing other than consciousness itself that survives the reduction. This reduction is, to reiterate, a reduction to the immediately given in experience. The significance of this reduction is that on the basis of this immediately given or intuited state of consciousness the constitution of the world is possible. In the natural attitude, Husserl claims, we may experience things and objects around us. But these things and objects, which are basically transcendent to our consciousness, should not stand as having any absolute link to us, because they come to us as the other; they are other than consciousness. Everything transcendent to our consciousness is put out of action in the course of the transcendental reduction. It is not the attitude of skepticism that denies the factual existence of something but rather a ground-breaking activity that captures what is immediately given in our consciousness so as to manifest a newly constituted world, of which the legitimacy of foundation cannot be asked any further. Thus, any phenomenologically meaningful things or objects are not something transcendent to our immediate consciousness, but something found and constituted within the confines of immediate consciousness—from what is given in our immediate consciousness. Husserl's starting-point may be the natural attitude, and the end-result may be the constituted world. But, in the course of the transcendental reduction, what is emphasized most is the role of pure consciousness, which is not only what is immediately given but at the same time the performer of the reduction itself.

Because the importance and emphasis are given to pure consciousness in this activity of the transcendental reduction and constitution, Husserl says that phenomenological meditations can best be carried out in the first person singular.[36] He even calls his philosophical enterprise 'a philosopher's personal affair', because one has to be attending to one's own immediate experience in conducting phenomenological meditations of the reduction to one's own consciousness and the deduction of the outward world within his inwardness of reduction.[37] Thus, Husserl's phenomenology becomes an 'egology', in which not only the performer of phenomenological research is the pure ego that cannot be done away with in any case, but also everything meaningful is found in this ego. As Husserl admits, this phenomenological enterprise is basically the ego's solipsistic philosophizing. For him, it is methodological solipsism that does not deny all other subjects or egos in the world. At the same time, it is indeed transcendental solipsism based on what Husserl calls pure egology. For Husserl, every conscious process is intentional in the sense that to be conscious is always to be conscious of something. In terms of this intentional characteristic, consciousness is capable of bearing its cogitatum within itself. Interestingly enough, this scheme of intentionality shows a similarity in its cognitive mechanism to Moore's distinction between the act of consciousness and the object of consciousness. As Moore's objects of consciousness are known to consciousness by the act of consciousness, so is Husserl's consciousness intentional by nature. Because of this intentional character, conscious process apprehends something transcendent to consciousness. Consciousness is always conscious of something. But in Husserl, what one sees in natural reflection does not mean much. Only when one has transcendental reflection, e.g., being conscious of one's conscious process, does one's ego stand as the transcendental ego over and above the naive empirical ego. Perceiving or experiencing the world already given as existing, the empirical ego remains in the natural reflection. It cannot be free from position-taking. Only by means of the transcendental ego, can we perceive that we are perceiving something, which enables us to see what we see purely as seen without any prejudice. This is possible through the transcendental ego's abstention from all positions already given. Thus, the transcendental ego is often described by Husserl as 'disinterested onlooker' or 'non-participant onlooker'.[38]

This division of the ego and the development of egology is one unique characteristic of Husserlian phenomenology directly related to what he calls transcendental solipsism. If the ego lies in one of those being experienced, it cannot be in the position of the pure subject that grasps all the experiences. The performer of intention itself cannot be the object of intention. Hence, the division of the ego secures the place of the transcendental ego in which all experiences can be entertained while it itself is not a part of experience. But the very phenomenological attitude of this transcendental ego is solipsistic because it cannot be other than the ego's own affair. When one attends to what is directly given in one's consciousness, what one does can only be one's own experiences. We can say that the ego, in this sense, is the totality of what it experiences or has experienced. In fact, this point is connected with Husserl's idea of the ego-constitution.

According to Husserl, the transcendental ego constitutes the objects around us. The transcendental ego is not just a passive point, its noetic activity plays an important role for the world constitution. Every object that is phenomenologically meaningful should be the outcome of constitution by the transcendental ego. Moreover, this transcendental ego constitutes not only the objects that have been bracketed but it also constitutes itself.[39] On the one hand, the ego has an intentional relation to objects in the constitution of the world. On the other hand, the ego is continuously constituting itself as existing. How, then, is this ego-constitution possible? According to Husserl, from every act generating from itself, the transcendental ego acquires a new abiding property, by which the ego can again be determined as a persisting ego. Even though each and every act of judgment passes by in the stream of consciousness and can even be canceled out, judgment results in an abiding property that the ego can repeatedly find as the ego's own. In other words, while the act of judgment vanishes, the outcome of the judgmental act itself persists. Whether the ego maintains this judgment or not, the very act of judgment gives the ego an abiding property as long as the judgment is accepted by the ego. Thus, the ego as the persisting ego is determined by this abiding habitus or state.[40] In this way Husserl can say "the transcendental ego is inseparable from the processes of making up his life,"[41] and "I continue to stand by my deed."[42] Not only does the ego play the role of the locus where the world of noemata becomes meaningful, but its own act of consciousness constitutes itself in turn. After all, what the ego experiences contributes to the constitution of the ego itself. Thus we could say that the ego is the totality of what it experiences or has experienced throughout its history.

Husserl in his late works attempts to show how to establish intersubjectivity from the solipsistic basis of the ego. Apart from the question of whether he is successful in his endeavor, what we have to focus our attention on is that Husserl's problem of solipsism originated from his emphasis on the idea of the attention to immediate experience or the immediately given. Since the most immediately given is my consciousness, solipsism necessarily results if the building process of every meaningful thing and knowledge is based on the given. This relation of solipsism to phenomenology is also implicit in Russell's case.

Placing himself in the tradition of British empiricism, Russell's concern is also about how we get certain knowledge from immediate experience. He wants to build our knowledge from the most certain data we immediately have. These data we immediately have in our experience, according to Russell, are sense-data. Being physical, sense-data are the sign of the existence of physical objects. They are neither sense-impressions nor mind-dependent ideas, but physical entities. Russell's ideal was that we could logically construct physical objects and therefore the physical world on the basis of a single person's sense-data. If we base our knowledge of the physical world on immediate data, solipsism is unavoidable, for sense-data are private. The idea of the logical construction from sense-data entails solipsism, because it is actual construction and not inference. To me, only my sense-data are known directly without any inference. Other people and their sense-data can only be known to me by inference based on certain sense-data. All that is certain to me is my existence and my sense-data.

To maintain this position that the knowledge of the physical or external world can be built up on a single person's sense-data or immediate experience, at least until 1916 Russell held that our cognition required a two-term relation between subject and object. There exists something given in my immediate experience. For this something to be known, there also has to be the subject which is the receiver of this thing. Strictly speaking, the subject (i.e., the ego) does not belong to what it experiences. The subject is simply a receiver of data, and there is no sign of the ego in the data itself.[43] Thus, Russell could show how we should describe what we know as follows:

When I look at my table and see a certain brown colour, what is quite certain at once is not '*I* am seeing a brown colour', but rather, 'a brown colour is being seen'.[44]

What Russell says is that, in reality, a sign of the subject is not found in the object itself and that the ownership often represented by 'I' or 'my' is an attribute given to the subject of experience. When we pay attention to immediate experience and try to explain cognition in terms of the relation between immediate data and the receiver of the data, solipsism is unavoidable, because all objects of experience should be meaningful and thus explained by attributing their ownership to the subject of experience. The most immediate and direct cognition can only be gained through the subject's having its own experience.

At the same time, in this scheme the status of the subject remains very difficult to deal with. Husserl, splitting up the ego into the transcendental ego and the empirical ego, tries to secure the ground for what he calls egology. But for Russell, a way out of this difficult problem is not entirely clear or satisfying, even to himself. In the Russellian scheme, the ego is a particular to which data are given. But Russell also believed, for the time being, that one can be acquainted with one's self. He was not quite sure whether one is directly aware of one's self alone, separately from all other elements that might accompany the self, for self-awareness is always the awareness of one's being aware of something. But still, Russell maintained that 'I', in addition to 'this' and 'that', is a logically proper name of a particular, which cannot be the case without direct acquaintance.[45] By the time *Mysticism and Logic* was published, however, Russell appeared to abandon holding 'I' as logically proper name.[46] What this could mean is that he no longer saw 'I' to be the object of acquaintance. Indeed, Russell no longer held that cognition is a two-term relation between the ego and sense-data, because he now saw the ego as logical construction rather than what we can have in immediate experience.

This solipsism implicit in our relation to immediate experience also appears in Wittgenstein. In Wittgenstein, as was argued earlier, no separate logical entity or categorial intuition are needed to figure out the logical structure of the world. Because of the built-in character of logical forms within objects, all that is needed in the whole logic and language is to experience objects. Hence, in Wittgenstein's phenomenology, the starting-point is not the transcendental reduction, and the end-result is not the constituted world. On the contrary, a phenomenologist's concern should be directed toward immediate experience, to be acquainted with objects and to figure out the logic of how the world is composed out of these objects. The basis in this scheme is to be acquainted with objects in immediate experience. Unlike Husserl's transcendental phenomenology, Wittgenstein's phenomenology does not involve any change of attitude, because Tractarian objects with logical forms in them are already the objects of purely direct experience with categorial articulation, the status of which can only be attained through the phenomenological reduction and constitutive activity in Husserl.

Regardless of the obvious differences in structure and terminology, there exist interesting parallels between Husserl and Wittgenstein. What plays a big part in Wittgenstein is the language to pick out immediately given objects. Picture theory explains how the elementary sentence of such language represents what goes on in the world. On the other hand, truth-function theory extends the picture idea to all the propositions of that language. Wittgenstein's scheme in this regard shows the reduction of everything cognitively meaningful to immediate experience.

This reduction to acquaintance foreshadows the presentation of solipsism in the later part of the *Tractatus* and some of his middle-period writings, for the emphasis on the idea of acquaintance necessarily puts the role of 'I' or the ego into the question. For Wittgenstein, the objects I am immediately acquainted with must be the objects of 'my' acquaintance. In other words, objects have to show up in my experience. They are 'my' objects in 'my' immediate experience. So, that the whole construction of the world depends upon 'my' objects is what Wittgenstein purports to show through his discussion of solipsism. Strictly speaking, objects in themselves do not show any sign of subject or ownership. But in phenomenological construction, attributing ownership to the subject is virtually unavoidable. Indeed, singling out an object in an egocentric way is one characteristic of phenomenology. For this reason, for Wittgenstein "the world is my world" and "what the solipsist means is quite correct"(*TLP*, 5.62). Wittgenstein's solipsism, too, is in this way virtually a natural consequence of his phenomenological position that gives primacy to immediate experience.

Wittgenstein also mentions the empirical subject and the metaphysical subject, drawing an interesting parallel with Husserl's empirical ego. According to Wittgenstein, however, there is no empirical subject in philosophy (5.631). There can only be the metaphysical subject, which is the limit of the world. The reason for this is also related to Moore's distinction between the act of consciousness and the object of consciousness. Considering that Wittgenstein's philosophical apprenticeship was undertaken in Cambridge, the possible influence, whether direct or indirect, from the ideas that belong to the tradition of British empiricism including Moore and Russell should not be underestimated. This Moorean cognitive scheme is one example showing that Wittgenstein's ideas are related to his British background.

Strictly speaking, we can differentiate three terms in the Moorean scheme of cognition, i.e., the ego (or consciousness), the act of consciousness, and the object of consciousness. The ego is the place where the objects of consciousness are given. The objects of consciousness exist outside the ego and are only known to the ego through the act of consciousness. So the role of the ego is merely having the objects of consciousness in its immediate experience. Before the initial act of consciousness takes place, objects still exist out there only without having any relation to the ego. By being given to the ego or by ego's having sensations, these objects become known to the ego as objects. Thus, in reality, there is no owner of the objects. But we cannot easily talk about the objects of experience without attributing their ownership to the ego in this scheme, because an object can be known as object through the ego's mental act. So the ego (or subject) is attributed as being the owner of experience. Wittgenstein basically follows this line of thought but not without his own problem. As hinted at in the *Tractatus,* one difficulty in this scheme is the status of the ego itself. For it to be cognitively meaningful, the ego itself has to be the object of experience. Since, for Wittgenstein, all objects have to be 'my' objects in 'my' immediate experience, the ego is the pole of experience, to which the objects of experiences are given. And this makes it impossible for the ego itself to become an object of experience. For this ego to be an object, there must exist another independent pole of reference to which this experience of the ego is given. Unlike Husserl, however, Wittgenstein does not develop this kind of egology. Instead, Wittgenstein denies the metaphysical subject as an object of experience. And as a result, this metaphysical subject is not part of experience, so that we cannot put it into words in any meaningful manner. For Wittgenstein, the subject is not only the totality of what it experiences, but at the same time a limit of the world. Hence, the subject itself cannot be an object of experience or a part of the world. It also sounds similar to Husserl's ego-constitution. But the difference is that for Wittgenstein, language is meaningful only when based on experience. 'I' is not a part of the world, and therefore, 'I' cannot properly be put into words.

I have argued so far the relation of solipsism to the phenomenological standpoint common to Husserl, Russell, and Wittgenstein. In relating solipsism to phenomenology we must be careful that this reduction to immediate experience or my experience does not imply or coincide with phenomenalism. That I confine the sphere of the cognitively meaningful to the objects I am acquainted with in my immediate experience might sound as if I have mind-dependent or subjective ideas. But, in no way are the objects of immediate acquaintance phenomenalistic. Because the solipsistic idea with regard to phenomenology is that the objects of immediate experience are the very base on which the ultimate analysis of the world can be carried out, this idea cannot be incorporated with phenomenalism that restricts our direct accessibility only to phenomena. What is important is that solipsistic ideas implicit in phenomenology do not entail that kind of restriction whatsoever. When we say that the objects of acquaintance is one's objects and the world is one's world, we do not mean that the objects or the world are dependent on one's ego. So the mind-dependent characteristic of phenomenalism cannot be the case in the discussion of the relation of solipsism to phenomenology. This point holds for Husserl, Russell, and Wittgenstein, and implicitly shows how the idea of solipsism goes hand-in-hand with that of phenomenology.

5. SOLIPSISM I

Earlier in this chapter, when arguing that Tractarian objects are the objects of immediate experience, I already touched on some of the early ideas of Wittgenstein's discussion on solipsism, which runs from 5.6 to 5.641 of the *Tractatus*. I argued that for Wittgenstein my language and my world share limits because my world is composed of the objects of my experience, which at the same time give the possibility of the meaning of my language in terms of their logical forms. This point directly leads us to what we may call a semantical solipsism, which Wittgenstein attempts to show in proposition 5.62.

> 5.62 The world is *my* world: this is manifest in the fact that the limits of *language* (of that language which alone I understand) mean the limits of *my* world.

The 'language' paraphrased in the parenthesis above means 'the only language that I understand'.[47] It is not a sort of private language that I alone understand, rather it is the only language that I conceptually have. Here we can see that Wittgenstein's attention is not directed toward the usual sense of solipsism which claims that only my mind exists or that the self and its states are the only possible objects of real knowledge. Rather, it is a semantical solipsism claiming that we are trapped in language in the sense that language is the totality of propositions that consists of names whose meaning is secured only by means of one's acquaintance with objects that bear logical forms. As the possibility of the entire logic of my language (and thus thought) is based on the logical forms of objects given in my experience, the limits of my language are the limits of my world. In other words, my language and my world share their limits. But what does it mean to say that my language and my world share their limits?

For Wittgenstein, objects are given to me in my experience. They are the objects of immediate experience. But this means that these objects are objects of my experience. When we think of the totality of these objects, it necessarily means the totality of my experience. And also, because the world is composed of objects, the world is nothing other than my world, which is the totality of these objects of my experience. Therefore, the world is my world (5.62). At the same time, my language also depends for the possibilities of its meaning on the logical forms of objects. Objects must be given to my experience in order that my language be possible, for the entire logic of my language is reduced to the logical forms of these objects. Therefore, my world and my language are related to each other in the sense that both are based on the givenness of objects. On the one hand, we can say that the totality of objects determines the limits of my world. Yet, we can also say that the totality of objects, because all the way in which objects can be combined is determined by the logical forms of objects, determines the limits of my language. In this way does Wittgenstein identify the limits of my language with the limits of my world. Indeed, the world as the totality of objects given in my experience would only result in drawing its limits in my language. By means of the given objects as logical atoms, I can make sense of my language. Hence, for Wittgenstein, not only is the world my world but my language and my world share their limits.

This point leads us to understand why for Wittgenstein the world and life are one (5.621) and I am my world (5.63). Since clearly my life is the totality of my experience[48] and the world is the totality of the objects of experience, it is not surprising that for Wittgenstein the world and life are one. From this, it also follows that I am my world. For Wittgenstein, all of this comes to mean I am my language, and thus is his discussion of solipsism semantically oriented. That I am the totality of experience or I am my world or I am my language all mean the same thing in this context and confirm again that the building-blocks of the world are the objects of my experience or that the logical atoms of language are given in my experience. Wittgenstein's points here are all based on the very nature of phenomenological objects. This phenomenological aspect of Wittgenstein's solipsism is properly expressed in 5.61.

5.61 Logic pervades the world: the limits of the world are also its limits.

This remark is a natural consequence of the view that the logical forms of objects determine the entire logic of our language and the way in which the world is arranged. It not only says that my language and my world share their limits but most importantly reveals that my entire conceptual scheme, i.e., the way I see the world and the way I use language to describe the world, is based on my immediate experience. The logic of my conceptual scheme entirely depends upon the givenness of phenomenological objects, so that I cannot escape from it. This is the point made by the remark "logic pervades the world." Wittgenstein continues to make this point in 5.61 as follows:

So we cannot say in logic, 'The world has this in it, and this, but not that'.

For that would appear to presuppose that we were excluding certain possibilities, and this cannot be the case, since it would require that logic should go beyond the limits of the world.

The logic that Wittgenstein mentions here refers to the way in which objects are combined together to construct the world. Indeed, this logic is prior to the question of "How are objects combined?" or "How is the world arranged?" (See 5.552). Thus, to say that the world has this in it and this but not that is what we cannot say in logic, because it is the question of what objects there are. Logic cannot answer this question, which goes beyond the limit of logic based on the givenness of objects. The starting-point of our logic is the givenness of objects in our experience, and thus we cannot ask beyond the point of givenness.

Wittgenstein says that the above idea provides the key to the problem of solipsism, and what the solipsist means is quite correct (5.62). Why is it so? Because my conceptual system is linguistically structured and my language is based on logic determined by the logical form of objects given to my experience. So my experience and my objects ultimately form language and the world. The world should be composed of my objects of experience, and language should also be meaningful from the logical forms of the given objects in my experience. In this sense, what solipsism says is quite right, and the underlying point of this solipsism is phenomenological. The limits of my language and my world are determined by the objects given to me, and I cannot escape from these limits.

What has been argued so far is roughly the first half of Wittgenstein's discussion of solipsism. He has focused on the limits shared by language and the world. But once Wittgenstein identifies 'my world' with 'I' in 5.63, he directs the discussion of solipsism to the treatment of the subject to which all the phenomenological objects are supposed to be given. So far, the nature of objects that 'I' experiences has been discussed. Next Wittgenstein attends to the status of 'I' as follows:

> 5.631 There is no such thing as the subject that thinks or entertains ideas.
>
> 5.632 The subject does not belong to the world: rather, it is a limit of the world.

We have seen that all the objects are the objects given in my experience; they are what I am acquainted with in my experience. The next problem then is the status of this 'I' which is the subject of acquaintance. Can I be acquainted with my 'self'? Can the subject itself become the object of immediate experience? What Wittgenstein says in 5.632 suggests that 'I' cannot be the object of immediate experience. For Wittgenstein, we cannot experience ego as an object. The subject is a metaphysical subject, which is a limit of the world

(5.641) and a point without extension (5.64). This ego is not what can be directly acquainted with in experience but rather is a point from which all objects make sense.

Then why does Wittgenstein say that we cannot experience the ego? Using a metaphor of the eye and the visual field, Wittgenstein explains that even though one can see everything in the visual field through one's own eyes, one cannot see one's own eyes in the visual field. For Wittgenstein, even saying "I see through my eyes" is not appropriate, because nothing in the visual field would allow one to infer that it is seen by one's eyes (5.633). Saying "I see through my eyes" is saying what cannot be said. It is nonsense because 'my eyes' are not found in the visual field. In other words, 'my eyes' are not part of the visual field but rather are the limit, which cannot be experienced in the visual field. 'My eyes' may function as the point from which the visual field is viewed and from which it is experienced. But as long as 'my eyes' function that way, 'my eyes' themselves cannot be the domain of experience. By the same token, the subject 'I' cannot be found in the domain of one's immediate experience. 'I' is not a part of experience. Hence, for Wittgenstein no empirical subject can be the object of experience. Only the metaphysical subject like the eyes in the visual field can exist. The metaphysical subject, showing a similarity with Husserl's transcendental ego, cannot itself be the object of experience but rather functions as the limit of the world.

Some remarks in the *Notebooks* are also helpful in making this point clear:

The I is not an object.[49]

I objectively confront every object. But not the I.[50]

As I argued earlier, Wittgenstein's objects are the objects of immediate experience. We name an object when it is presented to our immediate experience. To name an object, therefore, something must be given to us. Naming can be carried out by pointing to an object and then saying 'this' when we are acquainted with it. In Russell's case, 'this', 'that', and 'I' are logically proper names applied to what we are acquainted with in our experience. Even though he includes 'I' in the category of proper names, we find that Russell is not entirely satisfied with the status of 'I'.[51] But Wittgenstein has no hesitation regarding the status of 'I'. That Wittgenstein does not include 'I' in logically proper names means, as the above-quoted remarks show, that he does not think 'I' is an object with which to be acquainted. 'I' may be a point from which all the objects of immediate experience make sense. But 'I' itself is not an object with which to be acquainted, for 'I' is not a part of experience. In this sense, Wittgenstein's subject is not a part of the world. Rather, Wittgenstein conceives 'I' as the metaphysical subject, i.e., the ego that cannot be experienced but from which all experiences become meaningful. But that does not mean that those objects in the world are dependent on my ego (or the subject). Rather, what is important is that the objects are my objects and the world is my world. These objects have to be given, and all the ego has to do is to have these objects as its objects. The ego serves as the frame of reference for the objects to be meaningful. Thus, "the subject does not belong to the world; rather, it is the limit of the world" (5.632).

Hence, we cannot properly put this ego into words. It belongs to what cannot be said, for it does not belong to the domain of experience on which the logic of our language is based. The ego is not a part of the world, so there cannot be any mention of it (5.631). This point is repeated in Wittgenstein's middle-period writings, and it helps us understand his earlier position more clearly. In the *Philosophical Remarks* Wittgenstein says:

> The essential thing is that the representation of visual space is the representation of an object and contains no suggestion of a subject. (*PR*, sec. 71)

For Wittgenstein, 'I' is not an object with which to be acquainted or experienced. Thus, 'I' is not a proper name applied to the object with which one is acquainted. Nor is 'I' an owner of experience or sensation in the expressions of immediate experience. Wittgenstein says that 'I' in "I can see a red patch" is redundant, so that there can be another way of speaking the fact in which no mention of a personal pronoun like 'I' occurs. He means that 'I' is not an essential element in representing the facts. 'I' does not represent any objects with which to be acquainted. Nor are any objects in the world dependent on 'I'. Object and 'I' are distinct from each other. But objects are known to me by my experiencing of them. Thus, though we might express the idea as if 'I' is the owner of experience, in fact objects in themselves do not have any sign of ownership or subject.

What all this shows is that in the *Tractatus*, Wittgenstein's world is constructed out of the objects of my acquaintance, and thus it is phenomenological. But when the world is phenomenologically constructed, it becomes solipsistic in the sense that the subject plays the role of the frame of reference by which everything else becomes meaningful. In other words, in this framework, the world has to be based on my experience or my sense-data. What is important though, is that the subject of this phenomenological construction itself has no place in the constructed world, because the subject 'I' cannot be an object with which to be acquainted. Thus, we realize that Wittgenstein's phenomenological view regarding the objects of the *Tractatus* necessarily results in the discussion of solipsism as a consequence. In a very important way, Wittgenstein's solipsism in the *Tractatus* is a corollary to his basic phenomenological position.[52] When understood this way, why 'I' in "I can see a red patch" is redundant becomes clear. When I see a red patch, I say "I see a red patch." This form of expression misleads us to suppose that a subject to which the word 'I' is supposed to refer should exist. But it is simply wrong to suppose that kind of subject, because 'I' in "I see a red patch" does not stand for any object. 'I' cannot be an object with which to be acquainted in experience. In the *Philosophical Remarks*, Wittgenstein shows the relation of object to me, i.e., that an object is given to me, does not belong to the essence of object.[53] So my saying "I see a red patch" or "It is my sense-datum" does not mean that I am the owner of a certain sense-datum. Nor does it mean that there is a person to whom 'I' refers. 'I' does not play any role except for having the given objects. Other than that role, 'I' has no essential relation with objects. In other words, objects are simply given to me but are not dependent on my ego or my mind. Rather, the ego stands right in the limit of the world composed of these given objects.

These forms of expressions, which mislead us to see as if there was an essential relation between 'I' and the object of 'I', result from the phenomenological construction where 'I' serves as the frame of reference. In this sense, Anthony Kenny's view that no content can be given to 'I' in the 'I'-thought is right, although his approach is not phenomenologically-oriented.[54] Kenny's discussion of Wittgenstein's treatment of the subject 'I' is not motivated by the idea of the givenness of objects to one's immediate experience. Nevertheless, he maintains that 'I' is neither a proper name, a referring expression, nor the expression of ownership. Such sentences as "I am the thinker of these thoughts" or "I am this body" are philosophers' nonsense that often leads one to suppose the thinking substance as similar to Cartesian ego. The nonsense results from positing an idea about the subject to which one mistakenly tries to give a positive content. Yet this is the point Wittgenstein goes against. Thus he explicitly says that "the philosophical self is not the human being, not the human body, or the human soul" (5.641). It is the metaphysical subject or rather a metaphysical point with which all experiences are co-ordinated.

Wittgenstein's solipsism, as has been discussed, is a solipsism based on the phenomenological nature of logic and language. We can say it is solipsism implicit in phenomenology, which emphasizes the importance of the subject or the ego that entertains all the privilege of having pure or immediate experience. As I indicated in the previous section, we can interestingly compare Wittgenstein's view of the ego with Husserl's treatment of it. In Husserl's phenomenology a subject, putting everything out of action performs the transcendental reduction such that all the judgment regarding what has been assumed to exist thus far is suspended. In so doing, however, the subject itself survives the epoché as the transcendental ego. This transcendental ego does not belong to a usual sense of the world that has been put out of action. But instead, it is the constituting ego in which the sense-bestowing activity of noesis takes place to constitute the meant world. At the same time, Husserl does not deny the individually existing ego found in the individual's stream of consciousness, which he terms the empirical ego. The transcendental ego stands behind this empirical ego, and acts as the disinterested onlooker of all experience.

Therefore, for Husserl, as is the case for Wittgenstein, the subject is not and should not be a part of the world destined to be bracketed. But at the same time, the subject is the sole locus where the newly constituted world of noemata has meaning. In this Husserlian characteristic of the subject, we find a similarity to Wittgenstein's subject, which cannot be the object of experience. In a way, Husserl's transcendental ego functions analogously to Wittgenstein's eye of the visual field. Of course, the main problems of solipsism for the two philosophers are different. Wittgenstein's solipsism is linguistic in character, whereas Husserl's solipsism is concerned with intersubjectivity or the legitimacy of the transcendental ego with other egos in the sense that the transcendental reduction amounts to reduction of everything including other egos to the stream of my pure consciousness. But in both cases we find the characteristic of methodological solipsism that does not in fact doubt the real existence of other subjects. Also, the status of the subject in both cases is very similar in the sense that the subject has to play the role of the locus where all the other objects, and thus the world, gain meaning. For Wittgenstein, the subject is a point without extension with which reality is co-ordinated (5.64). On the other hand, for Husserl the subject is the very base that performs the phenomenological reduction and constitution of the world. In both cases, therefore, the subject is an overlapping point between the self and the world. In Wittgenstein, it is where my language and the world coincide. And in Husserl, it is where my consciousness and reality meet.

But the two views also have considerable differences. First of all, for Wittgenstein the fact that the ego is not a part of the world prevents him from speaking of the ego in any legitimate way. The ego belongs to what cannot be said, because its place is the limit of our logic. And also, Wittgenstein's metaphysical subject plays only a passive role. The subject is the extensionless point to which all the objects are given. On the other hand, Husserl's ego is not ineffable. The transcendental ego constitutes the objects around us, so that it is not just a passive point, but its noetic activity plays an important role for the world constitution. All objects that are phenomenologically meaningful should be the outcome of constitution by the transcendental ego.

6. PHENOMENOLOGY AND LOGICAL ANALYSIS

That Wittgenstein's objects presented in the *Tractatus* are objects of immediate experience analogous to Russellian objects of acquaintance provides an important clue to understanding one of the consistent and pervading themes of Wittgenstein's philosophy, namely phenomenology. These objects of immediate experience have a built-in feature of logical forms, which marks one of Wittgenstein's most decisive points of departure from Russell and eventually makes Wittgenstein's philosophy much different from Russell's.

Because logical forms are given together with objects, the experience we need in order to understand propositions is not the experience of a separate logical form. For Wittgenstein, all we need is to experience objects. Experiencing objects will secure the ground of understanding logical inference as well as having a meaningful language. There are no separate logical forms anywhere else other than within objects, and therefore these objects give not only the possibility of the logical atom of meaning but the entire construction of all possible logic in our language system. As I already discussed in section 2 of this chapter, Wittgenstein's rejection of separate logical forms contributes to the reduction of all logical forms to the logical forms of simple objects. On the other hand, Wittgenstein's truth-function theory amounts to providing the logical ground for extending the picture idea from elementary propositions to all propositions no matter how complex they may be.

As one of the several different theses of picture theory, the role of truth-function theory is to explain how complex propositions can be pictures without separate logical forms or logical constants. The very basic idea of picture theory is that an elementary proposition is a picture of reality by being an isomorphic copy of the corresponding state of affairs. The connecting link between language and reality in this relation is attained by one's experience of simple objects, which not only consist of states of affairs but provide logical forms automatically. One can directly experience the logical forms of objects, and that is all one is supposed to do to understand language. Other logical elements including the logical forms of complex propositions can and should be constructed out of these logical forms of simple objects.

For Wittgenstein, all we have is the logical forms of simple objects. Only with these logical forms must we be able to constitute all the forms of complex propositions. Because there do not exist separate complex logical forms or logical constants, all complex logical forms have to be built out of the logical forms of simple objects, so that all logical complexity can be explained by the direct experience of what is given to us. Russell, believing that complex logical forms were separate abstract entities, said that we actually experience those complex logical forms to understand complex propositions. But without complex logical forms as separate entities, Wittgenstein has to explain how we understand complex propositions as pictures of reality. In other words, one of Wittgenstein's tasks is to extend his picture idea, initially applied only to elementary propositions, to all propositions. His solution is the truth-function theory, the idea that all propositions other than elementary propositions can be constructed from elementary propositions, which he expresses in proposition 5 of the *Tractatus*.

> 5 A proposition is a truth-function of elementary propositions.

The way in which an elementary proposition becomes a picture of fact is by means of being a concatenation of names, which directly represent objects of acquaintance. And now, the way in which all other propositions become pictures of facts is by means of the truth-functional combination of elementary propositions. But this idea is fully understood only when supplemented with the actual way of attaining the proposed combination, as expressed in proposition 6 that all truth-functions are the conjunctions of negation of the given set of elementary propositions. This idea expresses nothing more than repeated applications of the Sheffer stroke operation using conjunction and negation. Wittgenstein clearly sees here that the operations of conjunction and negation could preserve the pictorial character of elementary propositions and extend it to all other propositions.

First of all, based on his fundamental idea that logical constants are not representatives (*TLP*, 4.0312), Wittgenstein attempts to replace all logical constants by the Sheffer stroke '|'. For instance, disjunct '$\vee$' can be eliminated by using '|', for we can say $(p|q)|(p|q)$ instead of $p \vee q$. Indeed, this replacement of all logical constants by the Sheffer stroke means that we can manage to do any logical operation by means of 'conjunction' and 'negation' only, as $p|q$ means $\sim p \& \sim q$. Having this idea in mind, Wittgenstein can assert 5.5.

> 5.5 Every truth-function is a result of successive applications to elementary propositions of the operation
>
> '$(-----T)(\xi, ...)$'.
>
> This operation negates all the propositions in the right-hand pair of brackets, and I call it the negation of those propositions.

The pair of brackets can be re-written according to 5.101, as $(FFFT)(p,q)$, which simply means $p|q$. And now Wittgenstein wants to represent this by a simpler notation $N(\overline{\xi})$, which amounts to the joint negation of the given set of elementary propositions, e.g., $\sim p \& \sim q \& \sim r \& \ldots$ (5.502). Then, the problem in this case is that, while he says the logical constants are not representatives, Wittgenstein employs 'conjunction' and 'negation' in his logical operations. Indeed, conjunction or negation is not the representative of fact but is rather used to extend the pictorial (or representative) character of elementary propositions to all other propositions. Now, how the pictorial character of elementary propositions is preserved by means of the operations of conjunction and negation should be explained.

That a conjunction of two elementary propositions makes one complex proposition representing one complex picture is easily understandable. Also, by the idea of polarity of sense, Wittgenstein can maintain that the negation of an elementary proposition represents the same state of affairs with the opposite sense.[55] What this means becomes clear when we refer to Wittgenstein's own remarks in the *Tractatus*.

4.0621 The propositions 'p' and '~p' have opposite sense, but there corresponds to them one and the same reality.

What Wittgenstein means by 'sense' of a proposition is its agreement or disagreement with the states of affairs. When a certain state of affairs exists, if what a proposition says is in agreement with the state of affairs, we call it true. If not, we call it false. (*TLP*, 2.21, 2.222) Thus, while *p* and *~p* have opposite senses, they have the same representational content, that is, "there corresponds to them one and the same reality." Since, for Wittgenstein, the logical constants are not representatives of facts, attaching '~' to a proposition *p* would not affect the fact corresponding to it. It is not that *p* represents a positive fact and *~p* a negative fact. Rather, *p* and *~p* represent one and the same fact. What we know when we understand a proposition is thus: "what is the case if the proposition is true, and what is the case if it is false."[56] In this way, negation is also used to preserve the pictorial (or representing) character but is not itself a representative. Therefore, when conjunction and negation are used in the successive applications of the Sheffer stroke operations, they indeed function as extending the pictorial character of elementary propositions to all other propositions.

Now, we can say that Wittgenstein's proposition 6 together with 5 is designed to show how the pictorial character of elementary propositions can be extended to all other propositions by means of truth-function represented as repeated applications of the Sheffer stroke operation. Because there are no separate complex logical forms or logical constants that enable us to understand complex propositions as in Russell's case, complex propositions should be understood as the results of operations with elementary propositions. This point is well made by Wittgenstein's own remarks:

5.234 Truth-functions of elementary propositions are results of operations with elementary propositions as bases. (These operations I call truth-operations.)

5.3 All propositions are results of truth-operations on elementary propositions. A truth-operation is the way in which a truth-function is produced out of elementary propositions.

Wittgenstein here shows how all the propositions of language are pictures of the world without separate complex logical forms or logical constants. So Wittgenstein can express his fundamental idea as follows:

> 4.0312 My fundamental idea is that the 'logical constants' are not representatives; that there can be no representatives of the *logic* of facts.

This and other similar remarks in 4.441 and 5.4 commonly reject logical objects from the inventory of the world. This point, incorporated with the built-in feature of logical forms within objects, throws light on the very unique and interesting character of Wittgenstein's logic.

For Wittgenstein, there are no complex logical forms or logical constants that enable us to make sense of complex propositions. Thus, Wittgenstein's logic is not the logic of complex propositions but is that of simple objects and names that constitute the atomic base of the entire world and language. In this logic all that is necessary is one's acquaintance with objects. The objects together with logical forms should be given in one's experience. Then, even the formation of a state of affairs among the given objects will be carried out by the determination of their own logical forms. All the necessary determination for the possible combination of objects to states of affairs is already built in the logical forms of objects. As Hintikka points out, "The formation of a state of affairs is not like gluing two cubes together, it is like fitting two jigsaw pieces together."[57] Wittgenstein's analogy that objects fit into one another like the links of a chain in a state of affairs (2.03) makes perfect sense in this regard. No additional 'logical glue' or other mechanism functions to combine objects together. What we need in addition to being acquainted with objects in Wittgenstein's logic is the truth-function theory that enables us to extend the pictorial character to all the propositions of our language and at the same time to reduce everything meaningful to the actual acquaintance with reality. All we have initially is names and objects. But all we have in our entire language and the world, too, is the combinations of those names and objects. So the consequence of Wittgenstein's jigsaw puzzle-like logic is enormous in the sense that it reflects Wittgenstein's whole project of the *Tractatus*, i.e., a codification of the entire conceptual framework based on the object of acquaintance, for what Wittgenstein purports to achieve through the truth-function theory is the reduction to acquaintance in our language. The fact that we understand and use the propositions of our language is ultimately based on the acquaintance of objects in our experience.

In the sense sketched above, therefore, truth-function theory is not entirely independent of picture theory, for by means of truth-function, complex propositions represent facts directly. The status of truth-function theory is better put as one ingredient of picture theory.

After all, truth-function theory in addition to the actual givenness of objects is the complete account of logic for Wittgenstein. Oddly enough, this is the whole story of Wittgenstein's logic, and no other element is needed. A very unique feature of this logic is that all we have is a picture that corresponds with facts. Hence, the logic is static in the sense that immediate comparison between picture and fact is possible.

Based on all these points, this logic is not like logic in its usual sense. By attending to immediate experience, according to this logic, we can figure out the entire conceptual scheme. Wittgenstein's logical analysis in this sense is identified with conceptual analysis in our language, and we can well again quote proposition 5.552 of the *Tractatus* describing the relation of 'experience' to logic.

> 5.552 The 'experience' that we need in order to understand logic is not that something or other is the state of things, but that something *is* : that, however, is *not* an experience.
> Logic is *prior* to every experience—that something *is so*.
> It is prior to the question 'How?', not prior to the question 'What?'

What experience gives us is the object together with its logical form. The entire possibility of *how* these logical forms determine the way in which objects are combined to build up the world is the question that can be asked in logic. But in the sense that this question should be asked in connection with 'experience', Wittgenstein's logical analysis coincides with phenomenology which amounts to reduction to immediate experience.

This radical conception of Wittgenstein's logic, however, faces a serious problem. For Wittgenstein, all we need in logic is truth-function theory. That should be the whole account of logic. If there are any other elements in logic such as the dependence between elementary propositions, it may not be compatible with the picture theory of complex propositions by means of truth-functional logic. In fact, the problem of color-incompatibility, which bothered Wittgenstein in his early middle period, worked as a force for him to give up the completeness of truth-function theory. This initial development of his early middle period clearly results in the introduction of a non truth-functional and thus non-pictorial element in the way in which language is linked to the world. Truth-functional logic functions as the means to guarantee the pictorial character for all the propositions of language, i.e., the way in which complex propositions are linked to the facts of the world. His abandonment of the completeness of this logic implies that Wittgenstein had to find another way in which this link is obtained. The seed of this problem is already immanent in the *Tractatus*, where Wittgenstein's famous remarks on color-incompatibility read as follows:

> 6.3751 For example, the simultaneous presence of two colors at the same place in the visual field is impossible, in fact logically impossible, since it is ruled out by the logical structure of colour.
>
> Let us think how this contradiction appears in physics: more or less as follows—a particle cannot have two velocities at the same time; that is to say, it cannot be in two places at the same time; that is to say, particles that are in different places at the same time cannot be identical.

These remarks might lead us to think that what Wittgenstein says with respect to the color-incompatibility issue is the physical structure of color. But, as Wittgenstein's repeated struggles with this issue in the early middle period shows, what he is really concerned with is to devise a suitable notation that would make color-incompatibility a mere logical truth of truth-functional logic. This point is also confirmed in Wittgenstein's 1929 paper "Some Remarks on Logical Form" and in the *Philosophical Remarks*. In the 1929 paper, Wittgenstein makes a surprising proposal and virtually gives up the completeness of truth-function theory in an attempt to solve the color-incompatibility problem.[58]

For example, if a proposition *p* stands for "A is green" and *q* stands for "A is red," the conjunction of *p* and *q* should not yield the normal truth-value as presented in the *Tractatus* (See Table 1),[59] because then, as the top line says, "A is green and A is red" will be true. But we know this conjunction is not true, and thus possibly one might suggest other truth-possibilities, as in Table 2, that would result in contradiction. But Wittgenstein argues that this is nonsense, and we all can agree. Surprisingly Wittgenstein's next step is to say the top line of *(TTT)* in Table 1 must disappear, because it represents an impossible combination of "A is green and A is red."

p	*q*	
T	*T*	*T*
T	*F*	*F*
F	*T*	*F*
F	*F*	*F*

(TABLE 1)

p	*q*	
T	*T*	*F*
T	*F*	*F*
F	*T*	*F*
F	*F*	*F*

(TABLE 2)

Why does Wittgenstein make this somewhat absurd remark in dealing with 'logical forms' as the title of the paper suggests? It becomes clear from this paper itself, as he explains what the problem with the color-incompatibility issue is. Here Wittgenstein really struggles with the difficulty in devising suitable notations that will coincide with the right logical forms for color ascription. Or to phrase it differently, Wittgenstein tries to find the mechanism of color phenomena with regard to our conceptual scheme. But he is unsuccessful within the Tractarian scheme of truth-functional logic. It is important for Wittgenstein to develop the correct symbolism to make it truth-functional tautology, for the complex proposition of '*p* and *q*', which is the outcome of truth-functional operation, should also be the picture that represents fact. Making an outrageous proposal to eliminate the top line of Table 1, Wittgenstein virtually gives up the completeness of truth-function theory. But his obsession to figure out the right logical form, and thus to get a correct notation, does not seem to disappear at the same time. Before getting down to the color issue in the 1929 paper, Wittgenstein elaborates on it as follows:

> The idea is to express in an appropriate symbolism what in ordinary language leads to endless misunderstandings. That is to say, when ordinary language disguises logical structure, where it allows the formation of pseudopropositions, where it uses one term in an infinity of different meanings, we must replace it by a symbolism which gives a clear picture of the logical structure, excludes pseudopropositions, and uses its term unambiguously.[60]

Interestingly enough, this quote echoes propositions 3.323-3.325 of the *Tractatus*, where Wittgenstein explains the way philosophical confusion arises and the way it can be cured.[61] However, remarkably, in the 1929 paper Wittgenstein makes it clear that the cure, i.e., developing the correct symbolism, can be provided by "inspecting the phenomena which we want to describe."[62] Wittgenstein clearly states here that his logical analysis consists of the inspection of phenomena, which is, as Wittgenstein therefore continues, an attempt to try to understand the logical multiplicity of phenomena.

Now, Wittgenstein ascribes the perplexity in color-incompatibility to our imperfection in symbolism. Thus, that we often reach an obvious nonsense as in Table 2 above is due to "a deficiency of our notation that it does not prevent the formation of such nonsensical construction, and a perfect notation will have to exclude such structure by definite rule of syntax."[63] From this we can say beyond doubt that the color-incompatibility of the *Tractatus* 6.3751 is indeed concerned with the correct notation that reflects the logical form of phenomena and not the physical structure of color in question.

In fact, Wittgenstein's earlier problem of color-incompatibility ultimately leads him to modify other major doctrines of the *Tractatus*. Just as the truth-functional theory has to prove incomplete, rejection of the independence thesis of elementary propositions is inevitable. What was presented in "Some Remarks on Logical Form" reappears in a more developed and definite form in the *Philosophical Remarks*. There, Wittgenstein once again shows his earlier perplexity with regard to correct symbolism in color ascription as "There is red here now" and "There is green here now." Wittgenstein says that the incompatibility of propositions like these should be recognized from the symbol alone.[64] Wittgenstein continues, "For since this [the symbol] includes the form of the objects, then the impossibility of '*f(r)&f(g)*' must show itself there, in this form." Now Wittgenstein realizes this is not the case. What he did not realize in the *Tractatus* was that there could be different logical forms (and thus different symbols) according to different cases. Wittgenstein thought of only one set of logical forms that applies to every mode of representation between language and the world. But now Wittgenstein articulates his new idea as follows:

> But the symbols do contain the form of colour and of space, and if, say, a letter designates now a colour, now a sound, it's a *different* symbol on the two occasions; and this shows in the fact that different syntactical rules hold for it.[65]

This new position indeed gives a new possibility for the explanation of an irritating problem, color-incompatibility. Wittgenstein is no longer willing to consider the proposition *f(g)&f(r)* nonsense or contradiction. On the contrary, he would say that "the '&' has a different meaning here, since '*x&y*' usually means *(TFFF)*; here, on the other hand, it means *(FFF)*."[66] Actually, this position was already presented in the 1929 paper. But it gets clearer with the elaboration that the word 'and' used in color words is different from the 'and' of logic. This position says much more is involved in the logic lying behind color than truth-functional logic.

One of the most immediate consequences of this change is that Wittgenstein's earlier plan to apply logic uniformly must be modified. We can no longer say that there is only one method of projection that relates language and the world but rather there are many different possible methods. As a consequence of this new view, Wittgenstein can say neither that complex propositions are made as the results of truth-functional operations, nor that elementary propositions are independent of each other. Thus, Wittgenstein no longer holds the idea that a single proposition corresponds to reality being the truth-function of elementary proposition. On the contrary, the proper picture should be this new idea that language as a whole corresponds to reality simultaneously. Wittgenstein states this idea not only in his writings but in his conversations with the members of the Vienna Circle.

> It isn't a proposition which I put against reality as a yardstick, it's a *system* of propositions.[67]

> I once wrote: 'A proposition is laid like a yardstick against reality. Only the outermost tips of the graduation marks touch the object to be measured'. I should now prefer to say: a *system of propositions* is laid like a yardstick against reality. . . . It's such a whole system which is compared with reality, not a single proposition. If, for instance, I say such and such a point in the visual field is *blue*, I not only know that, I also know that the point isn't green, isn't red, isn't yellow etc. I have simultaneously applied the whole colour scale. This is also the reason why a point can't have different colours simultaneously.[68]

Thus, for Wittgenstein we no longer compare each proposition with the corresponding state of affairs. Instead, we compare language with reality as a whole. As a result, the Tractarian thesis that elementary propositions are independent from one another loses its meaning. The above quoted remarks show not only what Wittgenstein's real concern including the color incompatibility issue was in the *Tractatus*, but what kind of change has happened in his early middle period.[69]

As Wittgenstein's philosophy around 1929 takes a new directions, so does his early notion of logical analysis begins to change. Because logic has more or less universal meaning easily identifiable with the truth-functional logic of the *Tractatus*, Wittgenstein's frequent use of the new term 'grammar' from his middle period on is worth noticing. In fact, 'grammar' replaces his earlier notion of logic. In other words, instead of emphasizing logical form as the only possible medium that bridges the relation between language and the world, Wittgenstein utilizes several different kinds of grammar or grammatical rules that govern color, sound, number, or electrical charges, respectively.[70] Wittgenstein shows how he sees the problem regarding his new important notion of grammar as follows:

The rules for 'and', 'or', 'not' etc., which I represented by means of the T-F notation, are *a part* of the grammar of these words, but not *the whole*.[71]

Thus, giving up the completeness of truth-functional theory does not mean any big shift in his basic concern, i.e., phenomenological analysis. How Wittgenstein views the problem of phenomeno-logic now has a wider focal point. In other words, the notion of logic is subsumed under the new notion of grammar, which dominates Wittgenstein's middle-period phenomenological enterprise as grammatical investigations.

CHAPTER III

THE LANGUAGES OF PHENOMENOLOGY

1. THE SHIFT OF LANGUAGE PARADIGMS

I argued in chapter 2 that the logic employed in the *Tractatus* is phenomenological in the sense that its possibility is based on the givenness of the objects of immediate experience. Consequently, I argued that Wittgenstein's language in the *Tractatus* is phenomenological language. However, Wittgenstein's writings in 1929 clearly show that a significant change regarding this view has taken place in his mind. The *Philosophical Remarks* based on his notebooks of 1929 and 1930 (MSS 105, 106, 107, and the first half of 108) is his first book-length work since the *Tractatus*. It begins with the following remarks:

> I do not now have phenomenological language, or 'primary language' as I used to call it, in my mind as my goal. I no longer hold it to be necessary. All that is possible and necessary is to separate what is essential from what is inessential in *our* language. (*PR*, sec. 1)

Thus has Wittgenstein's view on language changed. Wittgenstein postulated phenomenological language in the *Tractatus* to describe immediate experience. But now, as he returns to Cambridge to work as a professional philosopher, Wittgenstein rejects the necessity and even the possibility of phenomenological language. As I briefly mentioned in chapter 1, Wittgenstein now realizes that the only possible language is physicalistic language (i.e., everyday ordinary language). Now what we have to examine is Wittgenstein's problems around 1929 that caused his philosophy to change significantly. In other words, we have to spell out why Wittgenstein changed his views on language from the phenomenological system to the physical system. Answering this question surely cannot be easy as

Wittgenstein himself appears to undertake almost a year-long struggle with the problem as indicated in MSS 105-107.[1]

Wittgenstein's problem does not appear to be essentially new if we carefully consider the problems found in the *Tractatus*. There is no doubt that Wittgenstein is concerned with the link between language and the world in the *Tractatus*. The so-called picture theory is carefully designed to give an account of this bridge between propositions and what they represent. Strictly speaking, however, this picture idea applies only to elementary propositions. The role of extending this pictorial character of elementary propositions to all propositions of our language is performed by the truth-function theory. But the real and urgent problem Wittgenstein undoubtedly faced was the completeness of truth-functional logic. As we saw in the example of color-incompatibility, truth-functional logic is proven to be incomplete and thus not suitable to mediate between elementary propositions and complex propositions in applying the picture idea to all the other propositions of our language.

The first and the most immediate consequence of Wittgenstein's abandonment of the completeness of truth-function theory is that we can now no longer possibly compare all propositions of language with the facts of the world by direct means of truth-functional logic. But it appears that Wittgenstein does not give up the idea that propositions must be directly comparable with reality. This is shown among other concerns in the following:

> It would be a proposition that you could tell was true or false by looking at it. But how can you discover by inspecting the proposition or thought that it is true? The thought is surely something quite different from the state of affairs asserted by the proposition. (*PR*, sec. 44)

> You cannot compare a picture with reality, unless you can set it against it as a yardstick.

> You must be able to fit the proposition on to reality. (*PR*, sec. 43)

Thus, Wittgenstein still maintains that complex propositions have to be connected to reality. In other words, there should be a suitable connecting link between elementary propositions and complex propositions in some way other than truth-functional logic. As we saw in the previous chapter, abandonment of the completeness of truth-functional logic entails the introduction of non-truthfunctional elements in his logic such as the admission of the dependence between elementary propositions. Consequently, comparison between language and reality has to be done in such a way that a whole system of propositions is compared with reality rather than a single proposition is compared with a fact. The new way to mediate between elementary propositions and complex propositions, and thus the system of propositions and reality, is found in Wittgenstein's introduction to the notion of calculus that is operated with numbers and equations. As Hintikka points out, numbers that were merely exponents (or indices) of operation in the *Tractatus* become means of representation.[2] Wittgenstein takes the initial step for bringing numbers into the form of elementary propositions in his 1929 paper "Some Remarks on Logical Form."

> And here I wish to make my first definite remark on the logical analysis of actual phenomena: it is this, that for their representation numbers (rational and irrational) must enter into the structure of the atomic propositions themselves.[3]

The meaning of the quoted passage may not be clear at once. However, Wittgenstein elaborates in the *Philosophical Remarks* on what he wrote earlier and suggests that truth-functional logic is to be replaced by calculus.

> And if we say numbers are structures we mean that they must always be of a kind with what we use to represent them.
>
> I mean: numbers are what I represent in my language by number schemata.
>
> That is to say, I take (so to speak) the number schemata of the language as what I know, and say numbers are what these represent.
>
> This is what I once meant when I said, it is with the calculus [system of calculation] that numbers enter into logic. (*PR*, sec.107)

In other words, the logic of the *Tractatus* is now replaced by a calculus operated with number and equation. A different logic now performs the role of mediating between elementary propositions and complex propositions, a role performed by truth-functional logic in the *Tractatus*.

Like his earlier position, this new logic does not involve semantical elements for our language system. The logic of our language is now formal symbol operations by means of numbers and equations. Unlike his earlier position, however, conceiving propositions of our language as calculus operated with numbers and equations means that Wittgenstein conceives our language to be a dynamic system taking place in time and space. To compare language with reality, according to this idea, we have actually to do something with symbols. Language becomes a human activity, an idea that does not appear at all in the Tractarian scheme, which only requires static comparison between a single proposition and a fact. In fact, it is his first step toward the idea that language is a kind of game played by rules. Language is understood as the operations of calculus by rules.

The idea of static comparison in the *Tractatus* is put by Wittgenstein as follows:

> 4.05 Reality is compared with propositions.
>
> 4.06 A proposition can be true or false only in virtue of being a picture of reality.

Hence, a proposition (whether elementary or complex one) in the Tractarian scheme has only to become a picture of reality for comparison, which is static and does not require much ado. But when complex propositions are no longer viewed as pictures in Wittgenstein's abandonment of the completeness of truth-function theory and when a system of language as a whole is compared with reality, the method of comparison becomes a verification process by which we compare a proposition with what it represents. Thus does Wittgenstein say that "verification is the sense of a proposition."[4] The mediator of the verification process cannot be found anywhere else than in a proposition itself, and Wittgenstein comes to present the meaning of a proposition as the mediator of verification. He makes his point in the following:

> Isn't all that I mean: between the proposition and its verification there is no go-between negotiating this verification? (*PR*, sec. 56)

This verification idea is new in Wittgenstein's early middle period, and it subsequently opens up the way for his ideas of 'use', 'rules', and 'language-games', which are anticipated by the idea of calculus-like activity. But one of the most immediate consequences of the change illustrated thus far is Wittgenstein's rejection of phenomenological language.

Wittgenstein's new notion of language is a calculus-like activity taking place in time and space, which means symbol operations with pencil on paper that take physical means. Thus Wittgenstein can no longer hold the view that language is phenomenological; instead, our language is physical. Wittgenstein suggests this by way of how comparison between language and the world is carried out:

It's easy to understand that a ruler is and must be in the same space as the object measured by it.[5]

We must be able to *fit* a proposition *on to* reality /to put a proposition on top of reality/.[6]

Indeed, from the earlier period on, Wittgenstein takes very seriously that we must be able to compare a proposition with reality. Although some of the major ideas of the *Tractatus* change, Wittgenstein does not give up this idea of comparison between language and the world, but rather considers this his main idea (*Mein Hauptgedanke*).[7] Thus, his problem can be put this way: in order to be compared with one another, a proposition and a fact must not only share the logical form but also belong to the same space. But if we try to follow the Tractarian language paradigm, we cannot easily maintain this view. Wittgenstein immediately raises this problem:

If the world of data is timeless, how can we speak of it at all?

The stream of life, or the stream of the world, flows on and our propositions are so to speak verified only at instants.

Our propositions are only verified by the present. (*PR*, sec. 48)

Wittgenstein's problem here is that while they must do so, a proposition and what it depicts do not seem to belong to the same space. Why not? Because verification—i.e., the comparison between a proposition and a fact—takes place in time, while the fact we describe does not belong to time.[8] In the *Tractatus*, Wittgenstein's object is phenomenological, and thus language is phenomenological as well. And indeed these phenomenological objects do not belong to the temporal world, as Wittgenstein says:

> 6.3611 We cannot compare a process with 'the passage of time'—there is no such thing—but only with another process (such as the working of a chronometer).

Basically, the world of phenomena is atemporal. Being momentary, it is given in one's experience. But now in his early middle period, Wittgenstein realizes that verification only takes place in time, so now we can compare a proposition with reality only in time. This idea leads Wittgenstein to search for temporally persisting objects, which are indeed able to be verified in time. Our language, too, belongs to the temporal world, because it is verified by the present. In fact, Wittgenstein states that language belongs to the physical system.

> Language itself belongs to the second system. If I describe a language, I am essentially describing something that belongs to physics. (*PR*, sec. 68)

As I stated in chapter 1, what Wittgenstein means by the second system (i.e., secondary language) is physicalistic language as opposed to the first system indicating phenomenological language. Thus, declaring that language belongs to the second system marks Wittgenstein's serious departure from his earlier position and his shift of language paradigms from the phenomenological to the physicalistic. However, this change does not imply any change in the metaphysical status of objects. Of course, what phenomenological language describes are phenomenological states of affairs, which are momentary and given in one's immediate experience, while what physicalistic language speaks of is the physical world, which is not given immediately but persists throughout time and space. But Wittgenstein's point is not about the objects that exist or do not exist in the world or the metaphysical status of objects. Rather, he is concerned with verification, or as shown in his later writings of the middle period, with identification of the supposed object to be referred to by the symbols of our language—with the relation of language to reality. He once thought we could describe the momentarily changing world of immediate experience with ideal language, which reflects phenomeno-logic. The conclusion

Wittgenstein reaches in and around 1929 is that his earlier view is not possible. In order that we are able to compare propositions and reality—whatever the latter may be—the objects of reality must persist in time.

Colors are the examples of phenomenological objects for Wittgenstein. But the colors we have in immediate experience are not temporally persisting objects. All we have are constantly changing colors or color-images in color-space. Colors are in color-space as physical objects are in physical space. Wittgenstein comes to realize that the colors we immediately experience cannot be temporally persisting. What language can refer to should be able to be verified by calculus-like activities. And Wittgenstein realizes that there are no temporally persisting phenomenological objects found. Objects of immediate experience are in a constant flux like a river stream.[9] To be temporally persisting, an object has to be physical. Of course, here Wittgenstein does not posit two different kinds of objects. What he means is that even the objects given in immediate experience have to be referred to physically. We are able to describe objects in our immediate experience only physically.

Thus, ideal language that describes momentary objects is absurd; language has to be physicalistic. Wittgenstein no longer believes in the necessity of phenomenological language. And this is supported by Waismann's record of Wittgenstein's conversations as follows:

> I used to believe that there was the everyday language that we all usually spoke and a primary language that expressed what we really knew, namely phenomena. I also spoke of a first system and a second system. Now I wish to explain why I do not adhere to that conception any more.
>
> I think that essentially we have only one language, and that is our everyday language.[10]

That Wittgenstein rejects the necessity and possibility of phenomenological language and conceives physicalistic language to be the only possible language means that he has shifted language paradigms. Wittgenstein in no way also means to reject the objects of immediate experience or phenomenological objects. Wittgenstein does not reject his earlier position that there is the realm of immediate experience to be described. That Wittgenstein is still concerned with immediate experience and how to describe it is aptly shown in some of his own words:

All that is possible and necessary is to separate what is essential from what is inessential in *our* language.

That is, if we so to speak describe the class of languages which serve their purpose, then in so doing we have shown what is essential to them and given an immediate representation of immediate experience.

A recognition of what is essential and what inessential in our language if it is to represent, a recognition of which parts of our language are wheels turning idly, amounts to the construction of a phenomenological language. (*PR*, sec. 1)

Thus, Wittgenstein's rejection of phenomenological language by no means implies that there no longer exist any phenomenological problems. Instead, the above quote shows that it is still possible to conduct phenomenological analysis by figuring out what is essential and what is not essential in our ordinary physicalistic language.

2. SOLIPSISM II

After the rejection of phenomenological language in his middle-period writings, Wittgenstein still mentions phenomenological descriptions, which would describe what goes on in our immediate experience. A few examples can be listed as follows from the *Philosophical Remarks*:

Suppose I had such a good memory that I could remember all my sense impressions. In that case, there would, *prima facie*, be nothing to prevent me from describing them. This would be a biography. And why shouldn't I be able to leave anything hypothetical out of this description? (*PR*, sec. 67)

Isn't it clear that this would be the most immediate description we can possibly imagine? (*PR*, sec. 68)

If, now, phenomenological language isolates visual space and what goes on in it from everything else, how does it treat time? (*PR*, sec. 75)

Since these entries come after the rejection of phenomenological language, Wittgenstein's apparent acknowledgment of phenomenological descriptions or even phenomenological language sounds somewhat absurd. Yet what we should remember here is that phenomenological language only lost its sense as an independent and a primitive form of language. Wittgenstein's earlier belief was that there is phenomenological language that describes immediate experience. Realizing that language is a physical system in his middle period, however, Wittgenstein no longer defends his earlier belief. But, at the same time, he still has the problem of describing immediate experience, as he says that "the worst philosophical errors always arise when we try to apply our ordinary—physical—language in the area of the immediately given."[11]

Hence, there is a need to capitalize phenomenological descriptions (or phenomenological language) within the physicalistic language system as a means of describing immediate experience. The entries quoted above can be understood as operations of Wittgenstein's secondary phenomenological languages for the purpose of describing our immediate experience.[12] This secondary phenomenological language, which aims at hypothesis-free description, is not another kind of language but rather is reconstructed from within the physicalistic language system. Indeed for Wittgenstein, this amounts to achieving the same job as what the original phenomenological language would have done by figuring out which part of our language is essential and which part is turning idly.

That Wittgenstein employs secondary phenomenological language parasitic to physicalistic language system is possible because he is not concerned with the ontological status of what our language refers to. To be sure, Wittgenstein addresses phenomenological problems by primarily concentrating his attention on immediate experience or what is immediately given in one's experience. But, for Wittgenstein that does not mean there really exists a certain kind of world which may consist of immediately given data rather than another class of entities. He does not think the world consists of sense-data or physical objects. Wittgenstein's concern is more about language and how it operates when we use words like 'sense-data' or 'physical objects'. Wittgenstein says,

> And now don't think that the expression 'physical objects' is meant to distinguish one kind of object from another.[13]

Thus, attending to phenomenological problems is possible even after he rejects phenomenological language. Wittgenstein presents this matter in the context of solipsism in the *Blue Book*. How solipsism is related to phenomenological language is discussed in what follows. But first I want to point out that Wittgenstein considers phenomenological description or phenomenological language solipsistic if it is used independently of physicalistic languages. Wittgenstein believes however that phenomenological language has a role, i.e., to describe immediate experience insofar as it operates within the system of physicalistic languages. In other words, phenomenological descriptions are almost assimilated to the solipsistic way of speaking in the physicalistic language system.[14] I shall now examine how this assimilation happens.

First of all, Wittgenstein tries to characterize what a solipsist would claim:

> There is a temptation for me to say that only my experience is real: "I know that *I* see, hear, feel pains, etc., but not that anyone else does. I can't know this, because I am I and they are they."[15]

Also Wittgenstein says,

> In fact the solipsist asks: "How *can* we believe that the other has pain; what does it mean to believe this? How can the expression of such a supposition make sense?[16]

A solipsist insists that only his or her experiences are real. To Wittgenstein this sort of assertion has no place in his physicalistic language paradigm. But still, this might bother him. In normal use of language, it may not be a problem. However, our language has sentences containing the first person singular pronoun such as "I have a toothache," which Wittgenstein views to be fundamentally different from "He has a toothache." Wittgenstein's point seems already conclusive when he says,

> Now when the solipsist says that only his own experiences are real, it is no use answering him: "Why do you tell us this if you don't believe that we really hear it?"[17]

Wittgenstein implicitly shows that pushing the solipsistic position to assert its truth makes no sense. But, still in a way, Wittgenstein is prepared to admit that we use in our language some expressions in a solipsistic way. What he does throughout his argument is not to dismiss solipsistic expressions as nonsense but to show how their normal use becomes nonsense and to evaluate how we can use solipsistic expressions meaningfully in our language system.

Wittgenstein points out that our use of certain expressions is closely related to the way we identify objects. Thus, as mentioned earlier, it is not that there exists this or that class of entities and are called sense-data or physical objects. What is important is how we identify an object, i.e., according to the way we use language, we call this sense-datum and that physical object. This ontological non-commitment makes possible the use of solipsistic expressions, which are nothing other than phenomenological descriptions of immediate experience, i.e., how the world appears to us.

According to Wittgenstein, we have many different criteria for identifying objects. Notably he introduces two kinds of criteria in visual perception called 'the geometrical eye' and 'the physical eye'. Now a solipsist is a person who insists on using only the geometrical eye in identifying objects. The solipsist's method of identifying objects is a method of using a person's visual perspective as a pole of identification. On the other hand, a non-solipsist will also use criteria other than just the geometrical eye. Wittgenstein suggests here kinesthetic sensation as an additional method of using the physical eye as the pole of identification. But Wittgenstein's example is brief and rather obscure. Hintikka, using his own general terms for two different methods of identification, gives a comprehensive explanation in order that we can fully appreciate what Wittgenstein's introduction of the criteria really tries to explain. [18] Hintikka's corresponding term for the geometrical eye is the perspectival (ostensive) mode of identification, and, for the physical eye, the public mode of identification. In the perspectival mode, objects are identified or differentiated by either occupying the same location or not in one's visual space at the time of seeing. On the other hand, in the public mode objects are identified or differentiated by using publicly noticeable names or descriptions. These two modes are only for identification and in neither case talk about the two different classes of entities to which the modes might be used to refer. Exactly the same holds for Wittgenstein's case. With an analogy, Wittgenstein puts his relation between the two criteria:

> The grammar of the word 'geometrical eye' stands in the same relation to the grammar of the word 'physical eye' as the grammar of the expression 'the visual sense datum of a tree' to the grammar of the expression 'the physical tree'.[19]

Wittgenstein means that, when we talk about objects, we do not talk about ontologically different classes of entities such as sense-data or physical objects. The geometrical eye provides a way to speak of phenomenological objects; and the physical eye of physical objects. Only the methods of identification are important. In other words the ontological status of objects is not Wittgenstein's concern: to ask whether phenomenological objects really exist or not would be meaningless. What he purports to state is how we can possibly speak of immediate experience. And of course the geometrical eye provides us with the way in which we identify phenomenological objects and thus speak of our immediate experience.

The grammar of the geometrical eye is analogous to the logic Wittgenstein maintains in the *Tractatus*, which is ultimately the logical forms of objects we have when the objects are given to us. Here the way in which we recognize and identify objects is characterized by our pointing to an object in our visual field and saying "This." Such ostensive identification does not involve any physical means, whose employment necessarily brings in public character. Similarly, the grammar of the geometrical eye in Wittgenstein's middle period is what makes it possible for us to recognize and identify the objects of our immediate experience, e.g., sense-data, which momentarily appear to us. In fact, Wittgenstein himself does not give any explicit accounts of what exactly this grammar of the geometrical eye is. Perhaps he believed we cannot really talk about it. Indeed, if the grammar of the geometrical eye is analogous to the way we identify and name an object using the pointing gesture, we can say the nature of 'grammar' is ineffable, as Wittgenstein distinguishes 'showing' from 'saying' in the *Tractatus*.

Although Wittgenstein gives no full-fledged account of it, it is quite unmistakable that the grammar of the geometrical eye operates by means of only one criterion, which does not involve any physical means or public character. For instance, when a speaker in a lecture sees two members of the audience who are sitting side by side raise their hands to ask questions, the speaker can well use two different grammars to identify who is to ask the question. The grammar of the geometrical eye allows the speaker simply to say "You!" pointing at one of the two members with their hands raised in the speaker's visual field. Obviously, the speaker is clear about who he is pointing to. But the speaker's word "You!" with a simple pointing gesture may not be clear enough to the two people and other members of the audience to know who was really pointed to. This is using the identification method governed by the grammar of the geometrical eye. It makes perfect sense to the one who uses this method but to no one else. On the other hand, if the speaker points to one of the two members of the audience who raised their hands while saying "The one with glasses!" or "The one with the red jacket!" there will be no confusion by anyone, regarding who was intended to ask a question as long as the audience know what the words 'glasses' or 'red jacket' means. In this example, the words contain some physical attributions, and thus words like 'glasses' or 'red jacket' in fact help the speaker's solipsistic gesture of pointing not to be isolated from the audience. This identification method gives some sort of public means which removes possible confusion that might be present to anyone other than the speaker. This grammar makes it possible for us to recognize and identify continuously existing or temporally persisting physical objects. Thus, the grammar of the physical eye governs the way we identify physical objects, whereas that of the geometrical eye governs phenomenological objects such as sense-data. What is important here is that the same object can be identified differently depending on which grammar governs the identification and the expression of it. Hence, Wittgenstein would say that "those who say that a sense datum is a different kind of object from a physical object misunderstand the grammar of the word 'kind', just as those who say that a number is a different kind from a numeral."[20]

The Wittgensteinian problem in the above case is that even the way we express immediate experience cannot be carried out by using the geometrical eye only, for in that case the expression of immediate experience will inevitably be solipsistic. If the speaker tries to describe the speaker's immediate experience, i.e., noticing two hands in the audience, using the geometrical eye only, others may not understood the speaker's expressions as discussed. Wittgenstein's point is that in order to describe immediate experience meaningfully, it is necessary to bring in and rely on the physical eye.

This problem can be viewed from a different perspective, e.g., Wittgenstein's concern with the difference between the visual space and the physical space. The way we apply vocabulary of the physical space to the visual space could cause confusion.[21] For example, Wittgenstein says there is absolute position, absolute motion, and even absolute distance in the visual space, whereas position, motion, and distance are relative in the physical space.[22] Also, the visual space is not a space to which we can apply an exact measure. In the visual field, it would be nonsense to ask if we really see an exact circle as opposed to a very circle-like ellipse. Or it may well be possible not to distinguish a chiliagon from a circle with bare eyes only. In a strict sense, we have no perfect circle in the visual image. "An exact circle" is a word whose meaning must have originated from the talk of the Euclidean space where measurement using physical means can take place. In the visual space or any other realm of immediate experience, however, no physical means is involved as such. The visual space is governed by different grammar. Thus, Wittgenstein says:

> And so that a different sort of projection exists between the Euclidean circle and the circle perceived than one would naively suppose. (*PR*, sec. 212)

This point is calculated to show what his unelaborated analogy of the physical eye and the geometrical eye amounts to. Such words as 'exact', and 'precise' have their origin in physicalistic use in which measurement is involved. But we cannot measure visual images in our immediate experience. Nevertheless, we use those physicalistic vocabulary to express our immediate experience. We not only use 'exact' or 'precise', but also use 'inexact', 'blurred' or 'vague' to express what we see in our visual field. The confusion arises if we think we can apply these physicalistic words to immediate experience without any problem.

That we use those words which make more sense in physical measurement to describe immediate experience regardless of the danger of confusion is very important to the earlier example of identification methods and the grammar of the geometrical eye and of the physical eye. The point is, if we try not to use the physicalistic vocabulary at all in expressing our immediate experience, the result becomes very similar to what happened in the example of the speaker who only says "You!" with a pointing gesture. In other words, without relying on the physical eye, or without using the words governed by the grammar of the physical eye, we cannot successfully express to others what we immediately experience. If we insist on describing immediate experience without relying on the physical eye, we fall into solipsism.

A more general example in this regard is the case when we use words such as 'here', 'this', 'now', or 'I'. These words can be used in such ways to be governed either by the grammar of the geometrical eye or by that of the physical eye. Wittgenstein's problem is the confusion of these two different grammars, for we often think 'here' as the name of a place, 'this' as the name of a thing, 'now' as the name of a temporal point, and 'I' as the name of a person, which, in fact, they are not.[23] The name of a place, a thing, a temporal point, or a person all involves physical means of identification. But 'here', 'this', 'now', or 'I' is transparent to me whenever I use it to express my immediate experience. Wittgenstein's point is that these words can mean anything to the person who uses them when used in such a way that does not involve any physical means, i.e., when totally governed by the geometrical eye. On the other hand, those words mean nothing to others unless the words are aided by the physical eye. Language becomes solipsistic at this point. Thus, when we use 'this' to mean a thing, we would have to use it in such a way that it involves some sort of physical means, so that one can see 'this' (one thing) as opposed to 'that' (another thing). Similarly, we must use the word 'here' in such a way that others would be able to distinguish 'here' (one position) from 'there' (another position). Words like 'here' and 'this' become solipsistic when used in a way in which they are governed by the geometrical eye only. More generally, if we use an ostensive method of identification without employing any physical means, we fall into solipsism.

What is important here is that the way in which we identify phenomenological objects, or the way in which we apply the geometrical eye to our immediate experience, is expressed in a solipsistic way when put into words. In other words, the pure description of immediate experience is solipsistic because we are prone to rely on only one criterion when we cognize and identify immediate experience. Wittgenstein does not dismiss this solipsistic way of expression as meaningless. He acknowledges that a solipsist's employment of sense-datum language depending upon the geometrical eye alone makes sense as long as the use of sense-datum language is aided by the physical eye.

Wittgenstein's actual discussion about solipsism is given in an earlier page of the *Blue Book*, where a solipsist is described as a person who asks a necessity of notation other than ordinary language. In ordinary language, we use the expression "A has a toothache" (where A is not the speaker himself or herself). But the solipsist, who insists that only his or her experiences are real would try to use a notation whose rules will exclude the expression "A has a toothache" as meaningless.[24] This characterizes the solipsist position that insists the geometrical eye as the only legitimate criterion and excludes all others. But Wittgenstein embraces the solipsistic expressions as long as they are used in such a manner that they are dependent on a publicly identifiable physicalistic language system. This point is clearly shown by Wittgenstein's own example.

The man who says "only my pain is real," doesn't mean to say that he has found out by the common criteria—the criteria, i.e., which give our words their common meanings—that the others who said they had pains were cheating. But what he rebels against is the use of *this* expression in connection with *these* criteria. That is, he objects to using this word in the particular way in which it is commonly used.[25]

The solipsist's position has grounds only as one mode of identification. Solipsistic expressions governed by the geometrical eye cannot stand on their own as independent notations. If one insists that they can, then as Wittgenstein continues, one is not aware that one is objecting to a convention. Pushing the solipsist position and insisting that only his or her pain is real is a wrong move, and misses the point in his or her discussion. Thus, Wittgenstein puts the solipsist's point appropriately:

Sometimes the most satisfying expression of our solipsism seems to be this: "When anything is seen (really *seen*), it is I who see it."[26]

In this quote, the word 'I' does not mean the same as 'L.W.' even if the speaker of the word is L.W. The reason for this is all explained by the method of identification. The word 'I' makes sense to the geometrical eye (or the perspectival mode of identification), but the word 'L.W.' does not depend for its meaning on the geometrical eye. The meaning of the latter of course is given by the physical eye (or the public mode of identification) using several different public criteria. Wittgenstein himself asserts this point:

The word 'I' does not mean the same as 'L.W.' even if I am L.W., nor does it mean the same as the expression "the person who is now speaking." But that doesn't mean: that 'L.W.' and 'I' mean different things. All it means is that these words are different instruments in our language.[27]

Or rather, it means that these words are identified by two different criteria. One person (or thing) can be identified differently depending upon which criterion is applied. Saying 'L.W.' makes sense to the physical eye (or the public mode), while 'I' makes sense to the geometrical eye (or the perspectival mode.) Clearly, Wittgenstein's classification of 'L.W.' and "the person who is now speaking" on the one hand, and, 'I' on the other, reflects his earlier treatment of name and its relation to object. Strictly speaking, what Wittgenstein could accept as logically proper names are, following Russell, 'this' and 'that', which have no other means of definition except for ostension, because a name as a primitive sign cannot be dissected any further by means of definition (*TLP*, 3.26). In the *Tractatus*, a name serves to pick out what is immediately given in one's experience. What is directly given can only be picked out by logically proper names such as 'this' or 'that'. Thus, the existence of phenomenological objects could only be *shown*. From the middle period on, however, Wittgenstein appears to give up phenomenological language that describes immediate experience. But it is clear that he does not give up the entire way in which immediate experience is described. The very important role of 'showing' in the *Tractatus* revives in the middle period as a solipsistic way of speaking, which constitutes a part in our physicalistic language system. What differentiates the solipsistic way of speaking from physicalistic languages is the method of identification. When the geometrical eye is applied in the identification of objects, Wittgenstein would call it solipsistic. In the Tractarian language scheme, any objects given in my immediate experience are pointed to and picked out by means of logically proper names like 'this' or 'that'. In other words, when something is given in my experience, I point to the object and say 'this'. That is the only possible way of identification in this phenomenological language. Now, Wittgenstein calls it solipsistic because in that case only one criterion of identification, namely my own visual perspective, is applied to call the given object 'this'. According to Wittgenstein, that I point to the given object and say 'this' makes use of the visual eye only and lacks publicly identifiable means as my earlier example of a speaker's pointing to the audience. In this case, whether I point to the given object or not will make no difference, because in order that my pointing makes sense to others, I should be able to point to one thing as opposed to another, which will necessarily involve more than just my visual ostension. But when I apply the geometrical eye only, anything I point to will make sense to me. Wittgenstein illustrates his point in the *Blue Book* as follows:

> Now when in the solipsistic way I say "*This* is what's really seen," I point before me and it is essential that I point *visually*. If I pointed sideways or behind me—as it were, to things which I don't see—the pointing would in this case be meaningless to me; it would not be pointing in the sense in which I wish to point. But this means that when I point before me saying "this is what's really seen,"

although I make the gesture of pointing, I don't point to one thing as opposed to another.[28]

What Wittgenstein purports to show in the above passage is that in the solipsistic way of speaking, there is an inseparable connection between that which points and that which is pointed to, which makes the act of pointing meaningless. In order that the act of pointing makes sense, what is pointed at should be available to the physical eye. The object one wants to point to should be identified by more than one criterion. But when 'I' points to something and say "I see this," any connection between 'I' and the object cannot be established, because 'I' is not a publicly identifiable body or person. But of course, this 'I' is not something similar to Cartesian ego, either. Rather, we simply cannot give any positive content to Wittgenstein's 'I'. A similar example is given in "Notes for Lecture on 'Private Experience' and 'Sense Data'." Wittgenstein says,

"What is seen *I* see"(pointing to my body). I point at my geometrical eye, saying this. Or I point with closed eyes and touch my breast and feel it. In no case do I make a connection between what is seen and a person.[29]

But the point is that I don't establish a relation between a person and what is seen. All I do is that alternately I point in front of me and to myself.[30]

In this case, too, what is seen cannot be connected to 'my body' or any 'person' which is a publicly identifiable physical object. Rather, what is seen here is inseparably connected to the subject of experience, i.e., the geometrical eye.

Even though the analogy is not entirely revealing, for Wittgenstein a solipsist is analogous to a clock-maker who fastened the dial to the pointer of clock and made them turn round together.[31] In fact, that nature of solipsism results from the very mode of identification by the geometrical eye depending on one's visual perspective. As long as my visual perspective serves as the only pole of identification, my statement "Only this is really seen" becomes nothing but tautology. In this framework, anything I can say in the form of "I see this" or "This is seen" has to make sense and always be true, of course, to me. But within this solipsistic framework, I cannot successfully convince others of what I see, as the clock mentioned above will reveal nothing about the time. A case similar to this is the expression "I am here." To the speaker, "I am here" always makes sense and is also true. But in reality it makes no sense unless the speaker's voice is recognized and heard to come from a particular place of common space.[32] In other words, the expression "I am here" has to be available to the physical eye, by which we not only employ a visual means of cognition but also an auditory one, so that we can locate the speaker by figuring out which direction the voice comes from, together with visual identification of the location. But if the expression is not available to the physical eye, it is simply the same as saying "Here is here." In short, the solipsistic way of speaking serves to perform its role as long as it remains a part of the publicly identifiable physicalistic language system.

When the solipsist would exclude the expression like "He has a toothache" as meaningless, it is because the solipsist will only accept the perspectival mode of identification. If the solipsist also accepts the public mode, there will be no reason for the solipsist to assert that "Only my experience is real." As Wittgenstein himself suggests, we do not feel any difficulty in ordinary life with this solipsistic problem because in our physicalistic language system the solipsistic way of speaking plays a part. In other words, it is a matter of grammar. Even though they look similar in the form of expression, solipsistic expressions and non-solipsistic expressions are indeed governed by the different grammars. And now we can understand what Wittgenstein's phenomenological problem is. Allowing a solipsistic way of speaking, Wittgenstein makes it possible for himself to deal with what appears in one's direct experience. A solipsistic mode of speech plays a part in our language system. Or to put it differently, immediate experience cannot be meaningfully expressed without relying on the grammar of the physical eye.

In this way, his secondary phenomenological language employed for the purpose of hypothesis-free descriptions of immediate experience is connected to the acknowledgment of the solipsistic way of speaking in our ordinary language. Wittgenstein's secondary phenomenological language is supposed to be absolutely free of hypotheses and depends on the perspectival mode of identification;

i.e., it is governed by the grammar of the geometrical eye. Indeed, this language is absolutely free of hypotheses, because this way of identification and description is based on what one directly experiences and not mediated by any sort of inference, but it can only work within the physicalistic language system. The grammar of the geometrical eye cannot have primacy over that of the physical eye even in the description of immediate experience. Ontologically, there are no phenomenological objects to which this language might refer. But at the same time, it is not even correct to say what is described has a describing subject. This is what Wittgenstein means by his words "visual space has essentially no owner."[33] One might think that there is a subject which describes what appears. But being free of hypotheses, we cannot even presuppose a subject. Wittgenstein goes on to say that "Even the word 'visual space' is unsuitable for our purpose, since it contains an allusion to a sense organ which is as inessential to the space as it is to a book that it belongs to a particular person."[34]

The word 'visual space' indeed contains an allusion to the sense organ 'eye', which is physical, and the use of this word involves physicalistic grammar. It sounds as if the sense organ 'eye' or even 'I' the ego is the owner of the visual space. But the visual space is the space which totally depends on a perceiver's absolute visual experience, where the physical elements such as the 'eye' has initially no place to stand. It cannot be owned or possessed by 'I' either, which belongs to the same category as the words 'here', 'this', or 'now' which can be meaningful when used in such a way that we can distinguish 'this' from 'that' or 'here' from 'there', and so on. But without alluding to any word that involves the sense organ 'eye', this space cannot be expressed meaningfully. For if one insists not using words which involve physical attributions, any description of this space will be solipsistic. Without alluding to physical attributions, the perceiver of this space (the subject of experience) and this space itself cannot be individuated to others. What the perceiver describes cannot be conveyed to others. But by simply saying 'visual space' anyone can understand what it means, because its attribution to 'eye' enables others to distinguish this space from any other space such as Euclidean space.

Indeed, this point suggestively shows the nature of Wittgenstein's solipsism, and it is closely related to Wittgenstein's earlier discussion of the status of subject that cannot be the object of experience. If I try to build and describe the world on the basis of my own experience, solipsism is inevitable. At the same time, the status of the ego, which is the subject of all experiences, becomes very difficult to discuss for the ego itself has to be identified in one way or another. In the purely phenomenological description without any hypotheses, the ego cannot be identified by the physical eye. In fact, Wittgenstein's earlier notion of the metaphysical subject is revived here in the form of the geometrical eye. But the geometrical eye can function properly only insofar as it is operated within the physicalistic language system. Putting the solipsistic 'I' as a part of the physicalistic language system, Wittgenstein's language becomes non-solipsistic at last. So says Wittgenstein,

> "Solipsism // The conception of solipsism // does not stretch to games. The other can play chess as well as I.
>
> I.e., when we play a lang[uage] game we are on the same level."[35]

For Wittgenstein, the whole point of the solipsistic way of speaking is to seek the pure description of what appears to a perceiver without any hypotheses, and pointing out even the term 'visual space' is misleading. Indeed, Wittgenstein undertakes an effort to figure out the grammar of appearance.

Before closing this section, one thing to be discussed at this point is that Wittgenstein's discussion of solipsism shows an interesting similarity with Carnap's discussion of the methodological solipsism in the *Aufbau* as a basis of his logical construction of the world. As I discussed in chapter 1, section 2 above, Carnap relates his position to Husserlian withholding of judgment. At the beginning of his construction system, according to Carnap, we can speak neither of other subjects nor of the self, because the experiences simply must be taken as they occur. It is not until the logical construction has taken place that we can differentiate other subjects from the self. In this context Carnap says that the given does not have a subject. He also says that "*Egocentricity is not an original property of the basic elements*, of the given."[36] This sounds similar to Wittgenstein's remark that the word 'visual space' contains an allusion to a sense organ which is inessential to the space.

Carnap, unlike Wittgenstein, does not mention the way in which we might identify objects from this solipsistic basis. But noticing the danger of subjectivism, he attempts to give ground to secure the objective or intersubjective world by means of the definite descriptions which in nature deals with structural properties. In this case all that the statements care for is the form and not the content, and thus any individual subjective side no longer plays an important role. Carnap sees this intersubjective transformation to be the key characteristic of scientific statements;

> For science wants to speak about what is objective, and whatever does not belong to the structure but to the material (i.e., anything that can be pointed out in a concrete ostensive definition) is, in the final analysis, subjective.[37]

Even though Carnap does not make it explicitly clear, it is unmistakable that he, too, is concerned with two different ways of identifying objects in the solipsistic and the non-solipsistic frameworks of speaking. In the treatment of the given, Carnap, too, has to depend on ostensive definition, which is nothing other than using the geometrical eye in Wittgenstein's case. Carnap says that ostensive definition is subjective and solipsistic in the preceding quote. Moreover, he believes that he can gain intersubjectivity from the definite descriptions which only deal with the structural properties. This point is roughly analogous to Wittgenstein's physical eye (or the public mode of identification by Hintikka), because, as shown above, Wittgenstein's criteria are classified into the true proper names like 'this' or 'that', on the one hand, and the definite descriptions like 'L.W.' or 'the person who is now speaking', on the other. Therefore, it is worth noticing the fact that Carnap compares his methodological solipsism as his starting-point of construction to Husserlian transcendental reduction,[38] because the same methodological solipsism plays a crucial part of the phenomenological problem in Wittgenstein's case.

3. WITTGENSTEIN'S PHENOMENOLOGY OF TIME

I have argued thus far in this chapter that Wittgenstein, in his middle period, gives up some of the main ideas of the *Tractatus*. This change involves the rejection of phenomenological language employed in the *Tractatus* in favor of physicalistic language. One problem Wittgenstein has to take care of is the way we describe immediate experience, which was the job of phenomenological language. His solution, as discussed in the preceding section, is to capitalize on secondary phenomenological language within the system of physicalistic languages as a dependent notation. Indeed, we describe immediate experience successfully with our ordinary language. But the characteristic of those immediate descriptions of immediate experience is that the immediate descriptions are governed by a grammar different than ordinary grammar. This different grammar, which is phenomenological, or the geometrical eye as Wittgenstein would say, is allowed in our language, and we often use the expressions governed by this grammar. The only provision, though, is that the geometrical eye cannot be used as an independent criterion in our language use. What exercises the real power in our language operation is physicalistic grammar relying on public criteria.

Interestingly enough, we can find that this primacy problem between the two different grammars applies to Wittgenstein's notion of time, which is also closely related to the problem of solipsism. Unfortunately, we do not have much of Wittgenstein's written material about his view on time. But some fragmentary remarks given in his middle-period writings are good enough to shed light on this problem. Presumably, the first occasion where Wittgenstein elaborates his view on time is found in Moore's lecture notes.

> Later on he [Wittgenstein] made a distinction, as to the meaning of which I am not clear, between what he called 'memory-time' and what he called 'information-time', saying that in the former there is only earlier and later, not past and future, and that it has sense to say that I remember that which in 'information-time' is future.[39]

As Hintikka points out, the use of memory-time is to fix time-reference from the vantage point of the present moment, while that of information-time is to specify time-reference to the public by means of such objects as clocks or calendars.[40] What we can deduce from this distinction is that Wittgenstein's memory-time is the time concept we have in immediate experience, while information-time is the physicalistic time concept. According to Wittgenstein, in memory-time there is only earlier and later. This character alone makes memory-time the notion of time we have in immediate experience that does not include any publicly identifiable means of identification. It is very interesting to find that there is a similar idea in Russell's accounts of time in the following passage from his *Our Knowledge of the External World*.

> Immediate experience provides us with two time-relations among events: they may be simultaneous, or one may be earlier and the other later. These two are both part of the crude data; it is not the case that only the events are given, and their time-order is added by our subjective activity. The time order, within certain limits, is as much given as the events.[41]

What Russell wants to do is to logically construct time from our memory. The role of memory in this regard is very important because memory constitutes the basis of the actual construction rather than inference, which is less reliable in the Russellian scheme. This implies that Russell believes we can actually be acquainted with the events that happened in the past by means of memory.[42] That Russell indeed regards memory as one of the immediate sources of knowledge can be witnessed in the following examples:

> The first extension beyond sense-data to be considered is acquaintance by *memory*. It is obvious that we often remember what we have seen or heard or had otherwise present to our senses, and that in such cases we are still immediately aware of what we remember, in spite of the fact that it appears as past and not as present.[43]

> [A]s in most cognitive words, it is natural to say that I am acquainted with an object even at moments when it is not actually before my mind, provided it has been before my mind, and will be again whenever occasion arises.[44]

Here the sole means of deciding time-reference is our memory. By relying on the criteria of earlier and later, which involves memory, Russell says we can define a biography.[45] And this characteristic of Russell's idea of time-construction throws very interesting light on Wittgenstein's notion of time, as it appears that the early Wittgenstein more or less follows Russell's ideas of the theory of acquaintance. We even find that Wittgenstein uses the word 'biography', the meaning of which is not much different from Russell's.[46]

Nevertheless, what makes Wittgenstein's later notion of time different from Russell's notion is found in Wittgenstein's middle-period reaction to Russell's idea. Moore says in his lecture notes that Wittgenstein tries to show what is wrong with Russell's following statement: "Remembering, which occurs now, cannot possibly . . . prove that what is remembered occurred at some other time, because the world might have sprung into being five minutes ago, . . . full of acts of remembering which were, entirely misleading."[47] According to Moore, what Wittgenstein points out as Russell's mistake is that when Russell uses 'five minutes ago' as the time when the world might have sprung into being, his choice of 'five minutes ago' is arbitrary, and thus it says nothing quite different from "the world might have sprung into being two minutes ago," or even "the world might have begun to exist now."[48] Wittgenstein's criticism of Russell's statement is based on his distinction of memory-time and information-time, and to Wittgenstein, Russell only uses memory-time as a means of fixing time-reference. In fact, it is true that Russell only uses memory as a basis of time-relation that differentiates what happened in the past, e.g., 'five minutes ago', which is earlier than now. But what Wittgenstein takes to be a mistake in this is that the use of memory-time inevitably involves what can be called temporal solipsism. As a matter of fact, Russell is well aware of the formidable nature of solipsism, in this case experiential solipsism as distinguished from temporal solipsism, and says:

In one sense it must be admitted that we can never *prove* the existence of things other than ourselves and our experiences. . . . There is no logical impossibility in the supposition that the whole of life is a dream, in which we ourselves create all the objects that come before us.[49]

While realizing that we cannot produce a conclusive logical argument that will refute solipsism, Russell concludes that although solipsism is not logically impossible, there is no reason to suppose that it is true.[50] Thus, as indicated in Moore's lecture notes, Russell also believes that we can actually construct time from our memory, being well aware of the possibility of temporal solipsism.[51] When we rely on memory alone, it is logically possible that nothing exists except the present experience. Here Russell clearly admits the logical possibility of temporal solipsism, even though he does not use the word 'temporal solipsism'. However, as shown above, because Russell regards memory as an immediate source that actually enables us to be acquainted with what happened in the past, there is no reason for him to succumb to the possibility of temporal solipsism. While it is logically not impossible that nothing exists except the present experience, there is no reason to suppose this is actually the case, as it was possible for Russell to construct knowledge of the external things from the objects of acquaintance; now in the case of time, memory becomes the agent of extending the past experience to the present.

What we should turn to now is Wittgenstein's position. As has been emphasized, there is no reason to doubt Wittgenstein, who followed Russell's theory of acquaintance to a certain extent, also followed Russell's view on time. But as we witness in Wittgenstein's criticism of Russell's view in his lectures in 1930-33, Wittgenstein, in his middle period, no longer appears to be strongly influenced by Russell. It has been already pointed out that, in his early philosophy, Wittgenstein discussed solipsism, which is quite correct but is only inexpressible.[52] There, solipsism entailed because the world is my world in the sense that the world is constructed out of my sense-data given to me. Strictly speaking, this experiential solipsism coincides with temporal solipsism to Wittgenstein, of which Russell in his theory of acquaintance does not exclude the logical possibility. Indeed, Wittgenstein also believed that time had to be constructed by memory phenomenologically as the world could be constructed out of the building-blocks of my sense-data. On what ground can one claim that experiential solipsism coincides with temporal solipsism in the early Wittgenstein is not easy to spell out, as there is nothing explained in Wittgenstein's own text.[53] However, it is still possible to construct an argument for that on the basis of Wittgenstein's middle-period criticism of his earlier view, and this will be attempted later in this section with regard to the identification methods.

In the middle period, Wittgenstein begins to deviate from Russell's idea of acquaintance as he develops the notion of information-time, to which eventually he gives primacy as the basis of our time concept. Wittgenstein's deviation from Russell is shown in the above example where Wittgenstein criticizes Russell's mistake as the latter falls into temporal solipsism. From the Russellian point of view, as we have already seen, although there is the possibility of solipsism, experiential or temporal, we may well ignore the possibility, because the memory we have enables us to be acquainted with the event that happened in the past. Russell appears to have a strong belief in the direct acquaintance of the past solely by means of memory. In contrast, Wittgenstein in the early 30s appears to have different ideas. For Wittgenstein, the memory I have now should be my present memory and nothing else, so that I am not able to reach back to the past experience or to be acquainted with the past event by means of memory alone, which will necessarily result in temporal solipsism, unless it is aided by some other publicly verifiable reference of time. Thus, for him the statements "The world might have sprung into being five minutes ago" and "The world might have begun to exist *now*" do not make a big difference and can both be true, for both of them are based on one's memory that does not guarantee any direct acquaintance of the past which will verify one as true and the other false. Thus, for Wittgenstein the important role of specifying time-reference is assigned to information-time.

What is interesting here is that Wittgenstein does not dismiss memory-time as totally meaningless but only deprives it of absolute primacy. We can construct time solely by means of earlier and later based on memory. But when memory-time is used independently of information-time, temporal solipsism entails. To use one of Wittgenstein's favorite expressions, the grammar of memory-time is different from the grammar of information-time; the former is all right as long as it operates under the control of the latter. These two different grammars of time are also related to the two different methods of identification discussed in the previous section. According to Moore's lecture notes, Wittgenstein says,

> [W]hen we say "The clock is striking now," 'now' means 'the present of our memory-time', and cannot mean, e.g., 'at 6:07' because it has sense to say "It is 6:07 *now*."[54]

To rephrase this, when we say "The clock is striking now," 'now' cannot mean 'at 6:07' even if the statement was made at 6:07. At this point, one is irresistibly reminded of Wittgenstein's remarks that show the difference between the words governed by two different grammars, namely the geometrical eye and the physical eye. Although it was quoted in the previous section, it is worth quoting again to see the similarity to the case of time.

> The word 'I' does not mean the same as 'L.W.' even if I am L.W., nor does it mean the same as the expression "the person who is now speaking." But that doesn't mean: that 'L.W.' and 'I' mean different things. All it means is that these words are different instruments in our language.[55]

I argued that the word 'I' makes sense to the geometrical eye, and the word 'L.W.' to the physical eye. 'L.W.' and 'I' could mean the same thing (or person). But whether or not these two words mean the same thing is not Wittgenstein's immediate concern. What is important here, instead, is that the same thing, or person, can be identified differently depending on what kind of identification methods (i.e., grammar) is used.

While there is no wealth of material concerning this point in Wittgenstein's writings, it is unmistakable that we can also apply these two grammars to his notions of time, namely memory-time and information-time. Without doubt, memory-time is governed by the geometrical eye, and therefore phenomenological and solipsistic, whereas information-time is governed by the physical eye, and thus a physicalistic and non-solipsistic framework.

More evidence is found in the *Brown Book*. This time, Wittgenstein gives some examples regarding which is governed by the geometrical eye and which by the physical eye, respectively. The important thing here is that Wittgenstein places the word 'now', which is closely related to memory-time, on the same footing as 'I' or 'here'. According to him, "the function of 'now' is in no way comparable to the function of an expression like 'five o'clock', 'midday', 'the time when the sun sets', etc."[56] Wittgenstein here makes it clear that the two kinds of expressions are governed by the two different grammars, namely the geometrical eye one the one hand, and the physical eye on the other. When we compare the two statements "The sun sets at five o'clock" and "The sun is setting now," we are inclined to say that both 'now' and 'five o'clock' refer to points of time because of the similar forms of expressions. What Wittgenstein intends to point out is that we are deceived by the form of ordinary expressions which disguises the different grammar underlying the surface one. What Wittgenstein wants us to see is that some elements of our language are governed by what he calls the geometrical eye, which implies phenomenological grammar. The expressions governed by this grammar is solipsistic in nature. However, this does not mean at all that some expressions of our language are solipsistic, and therefore meaningless. All Wittgenstein wants to say by this is that we have to use these solipsistic expressions in such a way that they are also available to the physical eye so as to become non-solipsistic in the end. If we fail to do that, our use of those expressions falls into solipsism. And the way in which temporal solipsism is brought about is very similar to the way experiential solipsism is. Wittgenstein elaborates this point:

> One has been tempted to say that 'now' is the name of an instant of time, and this, of course, would be like saying that 'here' is the name of a place, 'this' the name of a thing, and 'I' the name of a man . . . . but remember that "This is short" without the pointing gesture and without the thing we are pointing to would be meaningless.—What can be compared with a name is not the word 'this' but, if you like, the symbol consisting of this word, the gesture, and the sample. We might say: Nothing is more characteristic of a proper name A than that we can use it in such a phrase as "This is A"; and it makes no sense to say "This is this" or "Now is Now" or "Here is here."[57]

Indeed, this passage indirectly shows what Wittgenstein's earlier view on time was like. The Tractarian objects were the objects of immediate acquaintance picked out by 'this' with a pointing gesture.

Since they were the objects given in my experience, those objects were my objects, and constituted the basis of experiential solipsism, and solipsistic phenomenological language that describes those objects of immediate experience. As his middle-period treatment of these solipsistic expressions shows, 'now' belongs to the same group of our linguistic elements as 'this' that is used for the direct acquaintance in our experience. But Wittgenstein makes it clear that the proper name 'this' cannot be understood as a means to give us the direct meaning unless it is used along with other publicly accessible criteria of identification. The way solipsism takes place is exactly when we fail to bring in this publicly accessible criteria. Hence, as 'this' leads to solipsism when used by the geometrical eye only as it was the case in the *Tractatus*, so does 'now' when used independently of the physical eye. But this time, it is temporal solipsism claiming that only my present experience is real. So we could say that the notion of time in the *Tractatus* is memory-time, by which Wittgenstein attempts to construct phenomenological time. As my world is constructed out of my sense-data, the whole concept of time, too, is constructed out of my memory. Thus, it is not strange that Wittgenstein says:

> Suppose I had such a good memory that I could remember all my sense impressions. In that case, there would, *prima facie*, be nothing to prevent me from describing them. This would be a biography.[58]

But since Wittgenstein no longer holds that memory-time is the absolute means in our time construction, his phenomenology of time only works within the physicalistic system where the word 'now' or our memory only plays a part. Hence, it is sensible to say that the notion of time that prevailed in the early Wittgenstein is memory-time which is not much different from Russell's idea of time-construction from memory, while information-time takes over the important role of time specification from the middle period on.

If Wittgenstein's notion of time in the *Tractatus* is memory-time, which has a strong kinship to Russell's idea of time-construction by means of memory, we can once again see an interesting parallel between Russellian-Wittgensteinian time-construction and Husserlian time-constitution, as we know that Husserl attempted to constitute time-consciousness by means of memory and expectation.

Husserl looks for the basis of time-consciousness from the givenness of the now-point, to which three unified, but distinguishable moments, namely sensation (or primal impression), primary memory (or retention), and expectation (or protention) belong. One reason that Husserl brings in the idea of retention in the now-point is that the now-point, while given, is not at rest. It undergoes constant change. It is in a constant flux, so that a 'now' immediately passes away or sinks back to the past.[59] Thus, in dealing with the perception of temporal objects, the problem of identity takes place. And the important role of 'retention' is to give us the possible means of identification. In a flow of time, before a 'now' sinks back into the past, what it has signified can remain present as 'has been' by means of memory, according to Husserl. What is more to this is we can even make the past once again the present. The now-point is actually given that constitutes what Husserl calls the original temporal field. When this given point no longer involves actual perception of new primal impressions, it becomes a past event. And it is the role of recollection (or secondary memory) that brings the past into the present. Husserl calls it 're-presentation' (*Vergegenwärtigung, Re-Präsentation*), as it is not actually perceived or given in immediate intuition but only appears as reproduced.[60]

Therefore, for Husserl, memory in general not only plays an important role in the identification of the temporal object when we perceive it in a flow of time, but it also contributes to re-presentation of the past. While primary memory (retention) is essential to the perception of ever-changing temporal objects, secondary memory (recollection) is crucial to the remembering, and thus, re-identification of the temporal objects.

Now, what we can get from Husserl's memory-based time-constitution is an interesting parallel with Russellian-Wittgensteinian way of memory-based time-construction. For Russell and the early Wittgenstein, memory plays a crucial role in enabling us to be acquainted with the past event. According to them, the criterion of identifying objects, in particular, temporal objects, entirely depends upon a perceiver's memory. Hence, they not only construct time by means of memory, but their construction also involves how we perceive and recognize beings. This idea of acquaintance by memory that gives the criterion of identification is in a sense similar to Husserl's idea of memory that has the power of re-presentation. Even though Husserl has a much more complicated theory of time-constitution, the motive of bringing in memory as a means of identification of temporal objects is strikingly similar to that of the Russellian-Wittgensteinian memory-time.[61]

It is also interesting to see this parallelism from the wider picture of Wittgenstein's philosophy. Wittgenstein eventually abandoned memory-time governed by the grammar of the geometrical eye as an independent means of time-construction. Interestingly enough, the time he began to give more emphasis to information-time almost coincides with the time he gave up phenomenological language. In other words, as Wittgenstein gave up phenomenological language, and thus the Tractarian phenomenological scheme, i.e., the construction of ideal language that describes immediate experience, he began to develop and emphasize the notion of information-time governed by the physical grammar. Moore's lecture notes were taken from Wittgenstein's lectures given in 1930-33, and Wittgenstein's remarks on time from the *Brown Book* quoted earlier were also made in 1934-35. The fact that Wittgenstein not only makes distinction between memory-time and information-time, but he also points out the danger of solipsism in memory-time in the writings after 1930 must have something to do with his philosophical transition.

Indeed, what we have to notice is that one of the crucial points Wittgenstein was tackling with regard to the possibility of phenomenological language involved the temporally persisting object. As was discussed earlier in this chapter, it was the temporally persisting object that Wittgenstein was looking for while considering the possibility of phenomenological language. This does not cause any serious trouble in Wittgenstein's early philosophy, where a single proposition is compared to a single state of affairs. But as the Tractarian scheme undergoes changes, a new idea of verification, which takes time, has been developed, and suddenly Wittgenstein is in need of temporally persisting objects to be referred to and verified in a new logical scheme. Unless there are temporally persisting objects, it would be a matter of course that this will lead to temporal solipsism. Wittgenstein struggles, but only to realize there is no temporally persisting objects that can be picked out in phenomenological language. Temporally persisting objects can only be identified by physicalistic grammar. It is thus a natural consequence that memory-time, or the phenomenological time concept, gives up its place for information-time. However, at the same time, this change means the phenomenological character of Wittgenstein's philosophy is getting weakened towards his later period, as his phenomenology operates only within the physicalistic system as a dependent geometrical grammar.

4. RULES, LANGUAGE-GAMES, AND PHENOMENOLOGY

In his middle-period writings, the language Wittgenstein employs is no longer the phenomenological language of the *Tractatus*. The only language Wittgenstein now has is physicalistic language, and teaching and learning of this language partly depend on ostension. In the *Tractatus*, the relation between language and the world was mediated by the pictorial character. Now in the middle period, Wittgenstein cautiously puts forward the idea of ostension as a medium of the language-world link. He indicates that ostensive definitions can give not only the meaning of words but also rules for the use of the word. In the *Tractatus*, the initial possibility of the meaning of our language was given by the role of the logical forms of objects. To be able to play this role of logical atoms, objects (i.e., the bearer of the logical forms) had to be given in our immediate experience as references of our basic symbols. Now Wittgenstein does not have this kind of logical atoms given in our experience. However, it is clear that Wittgenstein still utilizes some ideas evolved from his earlier notion of what makes our language work, or of the language-world link. How ostensive definitions play this important role can be shown by Wittgenstein's own words from the *Philosophical Grammar* :

> It may seem to us as if the other grammatical rules for a word had to follow from its ostensive definition; since after all an ostensive definition, e.g., "that is called 'red'" determines the meaning of the word 'red'. (*PG*, sec. 24)

> The correlation between objects and names is simply the one set up by a chart, by ostensive gestures and simultaneous uttering of the name, etc. It is a part of symbolism. Giving an object a name is essentially the same kind of thing as hanging a label on it. (*PG*, sec. 56)

The ideas presented in the quotes above are clearly different from Wittgenstein's later view in the *Philosophical Investigations*. In particular, the fact that Wittgenstein once held such a view that "Giving an object a name is essentially the same kind of thing as hanging a label on it" may come as a surprise, as this view is criticized by himself in the *Philosophical Investigations*.[62] But obviously, from the quoted passages, we see that Wittgenstein once believed that ostensive definitions could give meaning to words and the rules for their being used as well. Even his later view should not be misunderstood as a total rejection of the idea of ostensive definition. In his later view, it is language-games that constitute the foundation of semantics, in which ostensive definition plays a part. What Wittgenstein rejects is the idea that ostensive definition alone, without the public framework of language-games, can give meaning of words.

Contrary to his mature ideas, however, Wittgenstein's ostensive definition plays a very important role in bridging our language to the world in his middle period. That Wittgenstein believed ostensive definition could give rules for the use of words is very important, because it led him to conceive the rule to be the foundation of our language. The idea of ostensive definition as the bridge between language and the world is well presented in a few remarks of the *Philosophical Grammar* as follows:

> Of course an ostensive definition of a word sets up a connection between a word and 'a thing', and the purpose of this connection may be that the mechanism of which our language is a part should function in a certain way. . . . And it is the connection and not the effect which determines the meaning. (*PG*, sec. 138)

Wittgenstein says that ostensive definition sets up the language-world link and this link could determine the meaning in our language.

Wittgenstein sees ostensive teaching and learning of language to be one major source of meaning. Wittgenstein's point that we can teach the meaning of words ostensively reminds us of early Wittgenstein's near-identification of logic with phenomenology—hence, phenomeno-logic—that determines the entire logic and the possibility of meaning in immediate experience in the Tractarian language system. Wittgenstein's earlier point reappears as the form of ostensive definition. Ostensive definition is an immediate way of showing meaning (or even rules) to others. One can teach the meaning of a word to others by means of ostensive definition, which immediately connect the word to what it stands for. Hence, that Wittgenstein keeps holding the idea of ostensive definition as a major independent source of meaning in the middle period, and as a minor dependent source of meaning in the late period suggestively shows how Wittgenstein's phenomenological idea is deeply rooted, from the early period, and how it changed its shapes toward a mature view.

There is reasonable ground to assert this point. Wittgenstein appears to have once held that rules have phenomenological character. According to Hintikka, Wittgenstein once believed that "a rule for the use of a word could be obtained by attending to the experience I have when I understand the word."[63] What this implies is that it is possible to grasp rules by analyzing one's experience on one occasion. In other words, the rules are closely related to what we experience. Entries in the *Philosophical Grammar* support this:

> It can seem as if the rules of grammar are in a certain sense an unpacking of something we experience all at once when we use a word. (*PG*, sec. 12)

> We might call the recital of the rules on its own a criterion for understanding, or alternatively tests of use on their own.

> To understand is to grasp, to receive a particular impression from an object, to let it work on one.(*PG*, sec. 42)

Wittgenstein believed that rules are related to our experience in this sense.[64] And this remarkable point can be supported by Wittgenstein's own criticism of his earlier view as follows:

> Earlier I thought at one time that grammatical rules are an explication of what I experience on *one* occasion when I once use the word. They are as it were

> consequences or expressions of the properties which I momentarily experience when I understand the word.[65]

Even though this view that rules are phenomenological entities did not last long in Wittgenstein's philosophical journey, his longer entertaining idea of ostensive definition as a means to give rules and meaning can be understood in this context; that is, the phenomenological character of rules. For one who gives ostensive definition to another does the task of showing what he experiences in an immediate fashion.

Wittgenstein's emphasis on ostensive definition can also be seen in the *Blue Book*, where he gives it a privileged position as the only form of nonverbal definition. In fact, ostensive definition is not only the only form of nonverbal definition, but it is where "we seem to make a much more real step towards learning the meaning."[66] A few more examples regarding the status of ostensive definition which Wittgenstein assigns can be found throughout the *Blue Book*:

> If the definition explains the meaning of a word, surely it can't be essential that you should have heard the word before. It is the ostensive definition's business to *give* it a meaning.[67]

> If we were doubtful about how the sentence "King's College is on fire" can be a picture of King's College on fire, we need only ask ourselves: "How should we explain what the sentence means?" Such an explanation might consist of ostensive definitions. We should say, e.g., "this is King's College"(pointing to the building), "this is a fire"(pointing to a fire). This shews you the way in which words and things may be connected.[68]

Wittgenstein's emphasis on ostensive definition as one major source of meaning in our language means that he has formed and developed a new way of looking at our language since his rejection of phenomenological language in 1929. Our language is not like phenomenological language as it appears in the *Tractatus*. Rather, it is physicalistic. And the characteristic of this new language is that it involves calculus-like activity.[69] Wittgenstein conceives propositions of our language as calculus operated with symbols according to rules. He views our language as a physicalistic system taking place in time and space. There is no doubt that rules and rule-following become important in this newly developed view. Of course, Wittgenstein already has in mind the idea of language-games as early as the composition of the *Blue Book*, where he explicitly introduces the new term.[70] However, because Wittgenstein still relies on ostensive definition as one major source of meaning and rules, a predominant role as the foundation of semantics is not yet given to the idea of language-games. Language-games do not come before rules in the middle period. In fact, Wittgenstein still sees the rules for the use of words to be the fundamental ground of language-use. Wittgenstein's position of that time (roughly from 1933 to 1935) can be well characterized by his own words:

> You can't get behind the rules, because there isn't any behind.[71]

Thus, the rule is placed in a prominent position in Wittgenstein's treatment of language and how it is operated. In the later Wittgenstein, language-games legitimize all the rules by means of their being played.[72] It is clear that Wittgenstein comes to give language-games the foundational status in semantics. But in the *Blue Book*, rules still play a dominant role in our use of language. And this important role is mediated by ostensive definition.

Then, what made Wittgenstein change his mind? According to Hintikka and Hintikka, it is the question of the nature of rules.[73] At one point Wittgenstein thought rules were phenomenological entities. Later he rejected this view and brought up the idea of what it means to follow rules. In fact, the rule-following discussion occupies a lengthy part of the *Philosophical Investigations*, showing the weight of Wittgenstein's concern in his late philosophy. The initial treatment regarding this subject in the *Blue Book* does not coincide with his mature view, and it shows how Wittgenstein is developing his ideas toward the mature view.

In the *Philosophical Investigations*, where the rule-following discussion runs from sections 143 to 242, Wittgenstein argues that rules are ultimately determined by the language-games we play. The question of rule-following can be meaningfully asked only against the background of a particular language-game.[74] The ultimate bridge between language and reality is gained by language-games and not by rules or ostensive definitions. However, in his middle period,

Wittgenstein has not yet reached this conclusion. While he explains the ostensive teaching of the meaning of a word, Wittgenstein puts forward his view in the *Blue Book* as follows:

> The teaching may have supplied us with a rule which is itself involved in the processes of understanding, obeying, etc.; 'involved,' however, meaning that the expression of this rule forms part of these processes.[75]

Here Wittgenstein suggests that a rule is a kind of symbolic expression such as an algebraic formula or table that determines the course of rule-following or language-use. In other words, he suggests that ostensive teaching supplies us with a table which we actually make use of in understanding and following a rule. But attributing rule-following to applying symbolic expressions for it did not and could not satisfy Wittgenstein in the long run. The problem is, if ostensive definition gives a rule and this rule is something similar to algebraic formula, then there must be something accompanying the act of ostensive teaching and learning. Understanding and using language seem to involve some sort of experience that Wittgenstein may call queer. In this case, Wittgenstein should also give an account of what applying symbolic expression or formula or table means. A lengthy discussion of rule-following in the *Philosophical Investigations* includes some of the problems involved. In fact, the rule-following discussion is designed to de-emphasize the previously given role of experience as a source of meaning. This ultimately leads us up to the point where Wittgenstein explicitly shows the primacy of language-games over rules.

Wittgenstein's discussion of rule-following is rather obscure and sometimes not easy to follow. Nevertheless, what Wittgenstein attempts to show can be revealed by the critical examination of the negative attitude toward anything that might be possibly involved in understanding and following a rule. For example, what does it mean to say that following a rule is applying symbolic expressions? Do we grasp a formula in a flash and apply it, or do we need an interpretation to understand it? Wittgenstein says that we do sometimes have some sort of mental picture in mind. In this case, the applications of a rule come before one's mind (*PI*, sec. 140). But, soon he says that grasping a mental picture in a flash will not always work, because there can be what he calls abnormal case where this mental picture collides with application (*PI,* sec. 141). The problem is, to use an analogy, that even if one got the right algebraic formula, there can be more than one clear-cut application of the formula (*PI,* sec. 146). Thus, grasping a formula in a flash cannot serve the right explanation for what rule-following is, and Wittgenstein moves on to the examination of application.

Now, not just having a right formula but knowing how to go on, i.e., knowing the applications, is the criterion of understanding. But this will bring in another difficulty. Wittgenstein asks exactly when we can say we know the application. Do we know the application always, or only when we are actually thinking of the rule? (*PI,* sec. 148). Wittgenstein's negative remarks immediately follow. Understanding a word or rule is not a mental state so that we cannot pinpoint exactly when one begins to understand a word or to know a formula. A kind of mental experience may contribute to understanding. Wittgenstein does not deny that there is something going on in one's mind when one understands a word. But he is not interested in this sort of empirical matter. His point is, instead, that what is most important in understanding a word or rule is not this or that particular experience but involves human activity or practice (*PI,* secs. 149-155). At this point, Wittgenstein introduces another lengthy discussion of 'reading' and 'being guided' (*PI,* secs. 156-178). In this discussion, Wittgenstein needed to explain the case of a pupil who pretends to obey an order, instead of really obeying it; if rules are formally operable tables or formulas, one can follow the rules mechanically without understanding the actual meaning involved.[76] Thus, here too, the ultimate criterion for reading cannot be such behavior that a pupil reads out lines of words correctly. Rather, it has to be found in the practice and mastery of the game of reading. Neither can we know whether a pupil really reads or pretends to read nor pinpoint the exact point when a novice reader really begins to read. As long as we focus our attention on any possible mental experience of understanding and application, we are easily misled. Wittgenstein's conclusive point is already implied by emphasizing "'mastery' of a technique" (*PI*, sec. 150) and advising that we should "try not to think of understanding as a 'mental process' at all" (*PI*, sec. 154), which implies rule-following does not involve a special kind of experience and a word gets its meaning from the practice and mastery of language-games. Considering the interpretation of a rule will face the same fate. If interpretation is involved in rule-following, that interpretation will involve another interpretation, for one interpretation is not justified by itself.[77]

What appears to be the case is that Wittgenstein comes to realize the nature of rule and rule-following by referring to language-games. Rules do not belong to the phenomenological world and thus cannot be grasped by analyzing one's experiences on one occasion of rule-following. The experiences that may accompany rule-following have nothing to do with the determination of the way the rule is followed. It is not what he used to believe in the early 1930s. Rules do not determine the way in which they are followed, considering some of the problems outlined above. On the contrary, rules and rule-following are to be understood against the background of human activities called language-games. The rule and meaning of words are determined in the course of the entire practice of language-games. Wittgenstein, in fact, comments on the nature of a rule in the *Philosophical Investigations* as follows:

> Let me ask this: what has the expression of a rule—say a sign-post—got to do with my actions? What sort of connexion is there here?—Well, perhaps this one: I have been trained to react to this sign in a particular way, and now I do so react to it. (*PI*, sec. 198)

In this passage, Wittgenstein indeed reveals one of the most important points in his late thought. Using analogy, he assimilates a rule to a sign-post. If this point is clear, not much further imagination is necessary on our part to understand Wittgenstein's intention. What does it mean to say that we have a sign-post, and that we follow it? As Wittgenstein says, we are trained to behave in a certain way by the sign-post. This means, crucially, rules are not the foundation of semantics as they used to be in his middle period. There is a more fundamental ground than rules. Wittgenstein continues to explain what it is like.

I have further indicated that a person goes by a sign-post only in so far as there exists a regular use of sign-posts, a custom. (*PI*, sec. 198)

Wittgenstein means that a rule makes sense only if there has been a regular use of it. And this view obviously is the reverse of what he said in his earlier writings, if we remember that he used to believe that a rule can determine the way it is followed. Then, what does Wittgenstein mean by the regular use of rules, or customs? How can we characterize it? Wittgenstein's answer immediately follows:

To obey a rule, to make a report, to give an order, to play a game of chess, are *customs* (uses, institutions). (*PI*, sec. 199)

Wittgenstein never says that customs are language-games. But there is no doubt that 'to obey a rule', 'to make a report', 'to give an order', or 'to play a game of chess' is the defining characteristic of what Wittgenstein calls language-games. Thus, we can say Wittgenstein has reached the conclusion that rules (sign-posts) cannot have meaningful use unless there already exist language-games (customs). Perhaps one could develop and make use of one's own private rules. But the private rules will never be meaningful unless there already exists the practice of human activity called language-games. What Wittgenstein tries to show is, as Hintikka points out, language-games are conceptually prior to rules.[78] Language-games in the later Wittgenstein gain the most fundamental status in dealing with our language and language-use. Language-games are the rock-bottom of semantics. There can be no further thing to back up the idea of language-games. We can explain the meaning of a word, or how a rule is applied in terms of language-games, but not the other way round. Language-games themselves, therefore, do not need any other justification than simply showing the fact that we practice this and that language-game. There is enough evidence to support this point in the *Philosophical Investigations* :

If I have exhausted the justifications I have reached bedrock, and my spade is turned. Then I am inclined to say: "This is simply what I do." (*PI*, sec. 217)

Our mistake is to look for an explanation where we ought to look at what happens as a 'proto-phenomenon'. That is, where we ought to have said: *this language-game is played*. (*PI*, sec. 654)

The primitive language-game which children are taught needs no justification; attempts at justification need to be rejected. (*PI*, p. 200)[79]

Indeed, what we would really have to do is not "explaining a language-game by means of our experiences, but of noting a language-game.[80] That Wittgenstein gives a predominant status to language-games over rules implies he entirely gives up the possibility of the phenomenological character of rules and rule-following which he held one time in the middle period. In Wittgenstein's new perspective on language-games, the ostensive definition, too, has a meaningful part in our language-use only insofar as it works within the public framework of language-games. Customs do not constitute sign-posts. On the contrary, it is sign-posts that constitute what we generally call customs. Each sign-post is meaningfully understood against the background of an already established framework of customs. Analogously, when we name something, what makes that name meaningful is not a particular activity of ostensive definition, or a particular rule. Rather, it is the human activities of language-games in which ostensive teaching and the use of rules takes place.

When viewed from the wider context of Wittgenstein's entire philosophy, that language-games take over the job which ostensive definition has performed in the middle period has another significance. As was briefly indicated in the beginning of this section, the idea of 'showing', which is the mediating link between name and object in the *Tractatus*, is revived in the form of ostension and ostensive definition in the middle period. On the basis of the picture idea, the name-object relation constitutes one of the most important themes of Wittgenstein's philosophy, namely how we describe immediate experience in our language. The objects of immediate experience can only be picked out by applying names like 'this' or 'that', which is essentially an act of showing accompanied by the gesture of pointing. This idea of showing is present in the middle period in the form of ostension and ostensive definition, which perform the role of showing the meaning of a word in an immediate fashion. Indeed, ostension and ostensive definition work as the connecting link between language and the world. But as language-games take the supreme position in semantical ideas, language is related to the world by means of language-games. And all the phenomenological nature of immediate experience is absorbed in the larger framework of rule-governed human activities named language-games. Wittgenstein does not deny there are phenomenological realms in our experience. For him, there exist phenomenological problems through the very last moment of his philosophical life. The point is that even the way we show immediate experience has legitimacy on the basis of language-games. The entire practice of language-games constitutes not only the foundation of semantics but also the connecting link between language and the world. Reaching this conclusion from the middle-period battle, the phenomenological character of rules and rule-following is removed. What remains to be answered, then, is how the physicalistic language-games efficiently deal with the phenomenological realm of our experience, which has not been removed along with the rejection of phenomenological language from Wittgenstein's philosophical scene. This problem will be discussed in the next section about the private language argument.

5. SENSATIONS AND THE LANGUAGE OF SENSATIONS

Wittgenstein's main concern in his early philosophy is immediate experience and the way in which we describe it. In the *Tractatus*, Wittgenstein shows that we can pick out phenomenological objects given in our immediate experience by means of names. The task of naming, in fact, can be carried out by logically proper names such as 'this' or 'that' pointing to the given object. The phenomenological object is what immediately appears in the perceiver's experience. And there is no other way to pick out this simple object in this phenomenological scheme. As was mentioned earlier, the idea of pointing and naming was revived for some time in the middle period as ostension and ostensive definition. However, as language-games become the most fundamental ground of our language, the place of ostension or the phenomenological character of our language-use belongs to a parasitic part within the physicalistic language-games. 'This' or 'that' as logically proper names can no longer play the role of meaning-giving without being dependent on language-games. This idea applies not only to the everyday objects we see outside our minds but also to private experiences such as sensations we have internally. In the *Blue Book*, introducing intriguing ideas about the methods of identifying objects, Wittgenstein does away with the ontological claims about the external objects. Depending upon the way in which we identify objects, for instance, we call them either physical objects or sense-data. But still, internal sensations that one privately has are phenomenological, and the use of sensation-words seems to depend on one's own identification and re-identification. Wittgenstein still has to deal with the way we describe the immediate realm of private experience. So Wittgenstein's concern of immediate experience and the way we describe it from the early philosophy is still alive.

Apparently, there should be difficulty in dealing with private sensations and the expressions of them, for it seems that only the one who feels pain is accessible to his or her pain for certain and others can only guess because one's pain itself cannot be seen or otherwise felt by others. Regardless of this seeming difficulty, what Wittgenstein must do is to show that this phenomenological realm of internal experience has to be treated within the physicalistic language system—the only possible language system for Wittgenstein since 1929. Language system is physicalistic, and language-games we play in this system are based on public frameworks. Seemingly, the solipsistic nature of sensations would work against the way in which our sensation-words fit in this publicly accessible framework of our language and language-use. However, this worry or even difficulty can be resolved from the consequence of the discussion of rule-following examined in the previous section.

In the discussion of rule-following, Wittgenstein makes it clear that all the rules we make use of would be meaningless unless there already exists the practice of language-games. The very assumption that rules do not determine how the words are used but language-games determine how rules are followed excludes the possibility of private language. One can possibly develop one's own usage of words for one's private use. But one can make sense of the private usage because there already exists a regular practice of language-games. What we call rules and the meaning of words are the by-products of human activity named language-games. The primacy of language-games excludes the possibility of private language. What is really involved in this impossibility of private language is that even seemingly the most private and immediate realm of sensations, which can only be directly felt by the person who has the sensations, should be able to be expressed in our public language. Wittgenstein's conclusion in the *Philosophical Investigations* is, of course, that we can express the phenomenological realm of private experience in our language. Wittgenstein does not reject the existence of sensations in each person's mind. There is the private realm of sensations that each one of us immediately experiences. What he rejects is the solipsistic way of speaking about private experiences independent of a public framework that presupposes already existing language-games in practice. Wittgenstein's point is that we cannot ostensively point to sensation and give it a name directly, unless there is a mediating link by means of a public framework.

Perhaps, in the Tractarian language scheme, which is phenomenological as well as solipsistic in nature, name-giving to immediate experience, whether it is the experience of external objects like a table or of internal objects like a toothache, should be able to be carried out by simple naming-giving of 'this' or 'that' independently of any other mediating link. In that case, the ostensive name-giving itself is the mediating link. But now, Wittgenstein's criticism is that even that sort of simple name-giving, in particular the name-giving of the most immediate and private experience, is required to be mediated by the public framework of language-games. As long as we rely on this public framework, our talk of sensations makes sense. But if we use sensation-words independently of this framework, it becomes private language that is impossible in the language-game scheme. Indeed, this criticism of direct name-giving act of ostension is also against his earlier view presented in the *Philosophical Grammar*, where he says "giving an object a name is essentially the same kind of thing as hanging a label on it." (*PG*, sec. 56)

For Wittgenstein, the ostensive way of name-giving independent of language-games is no longer possible. In the case of private experience, if we try to set up the relation between a sensation-word and a sensation in such a way that one hangs a label on an object, the sensation-word would be meaningless, and the sensation itself would drop out as irrelevant. As was discussed in the previous section, for the later Wittgenstein, any semantical relation is established and secured by language-games. It is the basic link between language and the world. Practice of language-games is indeed the source of meaning in our language. In the middle period, it is clear that Wittgenstein once believed that name-giving is analogous to hanging a label on an object. But as he develops the idea that the language-game is the most fundamental ground of semantics, he comes to believe that name-giving, by means of direct ostensive definition, will not work. According to Hintikka and Hintikka, for the later Wittgenstein, an attempt to establish the semantical relation by means of ostensive definition is a wrong model. Focusing on Wittgenstein's pattern of using the words '*Bezeichnungen*' when not mediated by language-games and '*Namen*' when mediated by language-games, Hintikka and Hintikka call the former the '*Bezeichnung*' model, which is the principal target of Wittgenstein's criticism.[81] For example, in section 293 of the *Philosophical Investigations*, with the famous 'beetle-in-the-box' analogy, Wittgenstein appears to reject the idea that there exist private sensations. But when viewed carefully, it turns out that what Wittgenstein rejects is not the existence of private sensations but the wrong model on which one tries to set up the semantical relation. Since we cannot see other people's beetles, they might be different from each other, or someone might even have an empty box. But that is when we base ourselves on the wrong model in dealing with sensation-words. Wittgenstein says,

> That is to say: if we construe the grammar of the expression of sensation on the model of 'object and designation' [*Bezeichnung*] the object drops out of consideration as irrelevant. (*PI*, sec. 293)

Strictly speaking, the name-object relation (or any other semantical relations) cannot be set up without being dependent upon language-games in this case. Thus, Wittgenstein uses the word '*Bezeichnung*' to indicate the wrong model of 'object and designation', which is not based on language-games and would make 'beetle' irrelevant in consideration.

There are private sensations we feel everyday. We can indeed talk of pain as we name it. It is possible under the condition that we name pain within the framework of the language-game. And it will be impossible to name pain as hanging a label on it without the framework of the language-game. One of the core ideas regarding the so-called private language argument can be found in the section 244 of the *Philosophical Investigations*. There Wittgenstein points out the relation between sensations and sensation-words:

> How do words *refer* to sensations?—There doesn't seem to be any problem here; don't we talk about sensations every day, and give them names? (*PI*, sec. 244)

Wittgenstein does not go against the view that we can refer to sensations. It is a fact that we talk about our sensations in everyday life. Then what is the problem? Here Wittgenstein pays his attention to how the relation between sensations and their names are set up. Wittgenstein goes on to say,

> But how is the connexion between the name and the thing named set up? This question is the same as: how does a human being learn the meaning of the names of sensations?—of the word 'pain' for example. (*PI*, sec. 244)

Obviously, Wittgenstein's attention is directed toward the condition under which the names of sensations are connected to sensations as the first sentence of the above quote shows. But we have to be very careful of Wittgenstein's next move. In the second sentence, he rephrases the question and leads us to see the whole question of sensations and sensation-words from the broader perspective of language-teaching and language-learning. Even though he has not yet mentioned how the teaching and learning of sensation-words is possible, Wittgenstein's move implicitly shows that he is trying to embrace private experiences within his scheme of language-games that is in principle rule-governed human activities, which require practice. Teaching and learning of language cannot be carried out in someone's mind or imagination. Obeying a rule is a practice, and to think one is obeying a rule is not to obey a rule in practice (*PI*, sec. 202). So we could say that Wittgenstein's treatment of private experiences stands on the extended line of his discussion of rule-following. In the course of his rule-following discussion, Wittgenstein shows that the actual practice of rule-following constitutes language-games. The language-game we play is the most fundamental ground in language and language-use. Now, Wittgenstein attempts to show what consequence this language-game idea brings in the way that we express private experiences. The initial answer in this matter is that even when we deal with our sensations, we do the same thing as we do in the normal case of rule-following. The meaning of words and the rules to be followed are grounded on the human activities of language-teaching and language-learning. Without referring to the human activities called language-games, any talk of meaning and rules would be meaningless. The meaning of sensation-words is no exception. In the above quote, Wittgenstein clearly states that the way we get the meaning of the word 'pain' should be referred to the teaching and learning of sensation-words. Indeed, Wittgenstein is ready to say that sensation-words have meaning only against the background of the language-games of sensations in practice.

What Wittgenstein ultimately purports to show by the idea of the teaching and learning of sensation-words is of course not to reject each person's sensation, which is private to each other. On the contrary, Wittgenstein's point is that our expressions about private experiences become meaningful within the publicly accessible framework of language-games where the teaching and learning of the sensation-words can take place. Thus, Wittgenstein does not essentially and necessarily mean that we need a language community to have language-games. For the same reason, Wittgenstein in principle does not deny private language that can be 'used' by only one person.[82] Thus, it is not because we cannot identify any criteria for our inner sensations that Wittgenstein rejects private language. In fact, Wittgenstein takes the example of a diary-keeper who fails to identify the recurrence of a certain sensation (*PI*, sec. 258). The diary-keeper might try to concentrate his attention on the sensation. But the diary-keeper cannot establish any connection between the name of the sensation and the sensation itself. From sections 258 to 269 of the *Philosophical Investigations*, with different thought-experiments, Wittgenstein repeatedly shows that any attempt to look for the criterion for identification of sensations by introspective means such as concentration or memory is bound to fail. However, that we cannot have the criterion for identification of sensations is not the ground on which Wittgenstein argues against the possibility of private language. It is not because we cannot trust our memory in recalling what our earlier pain was like; indeed we may be able to establish a reliable criterion with a help of mechanical device (*PI*, sec. 270). So it is not because memory is fallible and thus can deceive one's judgment that Wittgenstein excludes memory as a criterion for identification. At bottom, whether memory is always correct or not does not matter at all because Wittgenstein here is not concerned with each and every actual empirical case where memory is involved. He simply points out that memory cannot be correlated to an independent public framework. It is logically impossible that relying on memory in re-identifying sensations becomes a move in the language-game of sensations. Any attempt to set up the relation between the name of a sensation and the sensation itself by means of criteria other than the entire practice of the language-game is ill-founded. No criteria are needed except for the role of the sensation-words in the entire language-game. The way we learn the meaning of sensation-words is to take part in the same practice as others do and get to know the way the words are used in the practice of the game. There indeed is no criterion for the identification of sensations. Instead, paying attention to our language-game and the way we play it will resolve the problem of identification of sensations. Other than the language-game we simply play, can there be no further justification or fundamental ground to mediate the relation between sensations and the expressions of sensations. Looking for the criteria for the identification of sensations is analogous to a wheel turning idly (*PI*, sec. 271) or the ornament of a machine which is not connected with the mechanism at all (*PI*, sec. 270).

Of course, we can think of a possible situation in which a child is pretending to have a toothache, as in the earlier example where a novice reader who pretends to read. If the child is a good actor, others would never know whether the child is really having a toothache or not. Thus, we might be in need of some criteria to know whether there is real pain in the child's tooth. But as I have argued, and as Wittgenstein assures us, this is not a problem of the language-game of sensations itself. Of course, one can pretend one' pain, and we can doubt others' pain behavior. In that case, we might also need some sort of criteria to clear our doubt. But pretending and doubting become significant games only after we have regular practice of more basic games of pain. In the case of pain-game, the sensations we have are directly reflected on our sensation-words. There exists a direct interface between language and the world in this case. Here we do not need any criterion. The fact that we are playing the language-game is itself the criterion we need. The situation in which one pretends one's pain, or whether one doubts someone else's pain comes later than this original language-game of pain. Wittgenstein clearly comments on this point.

I turn to stone and my pain goes on.—Suppose I were in error and it was no longer *pain*?—But I can't be in error here; it means nothing to doubt whether I am in pain!—That means: if anyone said "I do not know if what I have got is a pain or something else," we should think something like, he does not know what the English word 'pain' means; and we should explain it to him.—how? Perhaps by means of gestures, or by picking him with a pin and saying: "See, that's what pain is!" This explanation, like any other, he might understand right, wrong, or not at all. And he will shew which he does by his use of the word, in this as in other cases. (*PI*, sec. 288)

In the original language-game of pain, there is no room for making an error of using the word 'pain'. If one is trained to use the word 'pain' by practicing the language-game of pain, one should be a perfect user of the word. And other participants of the same language-game should also be perfect users. Problems of pretending or doubting do not arise in this stage. But it is possible that people behave many different ways, and indeed pretend their pain. And Wittgenstein now explains this case.

That expression of doubt has no place in the language-game; but if we cut out human behaviour, which is the expression of sensation, it looks as if I might legitimately begin to doubt afresh. My temptation to say that one might take a sensation for something other than what it is arises from this: if I assume the abrogation of the normal language-game with the expression of a sensation, I need a criterion of identity for the sensation; and then the possibility of error also exists. (*PI*, sec. 288)

As Hintikka does, we may term 'the normal language-game' a primary language-game, and 'the abrogation of that game' a secondary language-game.[83] The language-game of sensations is indeed a primary language-game where language-use and reality have a direct relation. In this case, language-game is the link between language and the world. No other elements can come in-between in this game. We do not have to look for criteria for meaning, because playing this game itself gives us meaning. But once we master this game, we can develop a sort of applied games such as pretending and doubting. And these are secondary language-games, because we indeed build these games on the primary language-game.

As Wittgenstein admits it in the above quote, when one pretends one's pain, we need a criterion of identity for the sensation. But in the primary language-game of sensations, nothing can serve as the mediator between sensations and the expressions of sensations other than the language-game. One should only be concerned with the way the signs of sensations are employed in the language-game of sensations. Just as ostensive definition cannot work independently of already established language-games, so would the sensation-words not work without reference to the publicly available language framework. This way, even the most direct, immediate, and private experience can be expressed in public language-games. Going back to section 244 of the *Philosophical Investigations* quoted above, we now witness how Wittgenstein explains the way that our language of private experience can become public as follows:

[H]ow does a human being learn the meaning of the names of sensations?—of the word 'pain' for example. Here is one possibility: words are connected with the primitive, the natural, expressions of the sensation and used in their place. A child has hurt himself and he cries; and then adults talk to him and teach him exclamations and, later, sentences. They teach the child new pain-behaviour.

"So you are saying that the word 'pain' really means crying?"—On the contrary: the verbal expression of pain replaces crying and does not describe it. (*PI*, sec. 244)

The previously quoted sentences of section 244 indicate that we need a public framework of language in order to be able to speak of private experience. The ground that the name of a sensation can function as a name—for instance, the word 'toothache' as the name of a toothache—is secured only through the mediation of a public framework. In other words, in order that we can refer to sensations, we need language-games. Most importantly, thus, the meaning of sensation-words comes from the language-game of sensations. What Wittgenstein illustrates in the above example is how it is possible. One distinct feature of the example in which a child learns the meaning of sensation-words is Wittgenstein's view that there is inseparable connection between sensations and the behavior that accompanies sensations. When one feels pain, the behavior of crying, exclaiming, grimacing, or others may accompany the pain as natural expressions of pain. Thus, the usual internal sensation of pain is outwardly manifested by pain-behavior.[84]

One is tempted to view, at this point, Wittgenstein's private language argument as an argument in support of behaviorism, which holds that sensation-words do not refer to the inner experiences or properties but to the overt and publicly observable behavior or dispositions to behave in certain ways. But the behavioristic interpretation of the private language argument is a misunderstanding based on the false assumption that Wittgenstein rejects the existence of private experience. As was mentioned in the beetle case, however, sensations private to each person do not drop out of consideration as irrelevant. Instead, by means of their external manifestation of sensation-behavior, sensations come to take part in our language-games. Hence, that there are outward sensation-behavior that accompanies internal sensations should not be understood as a point in support of behaviorism. It, in fact, plays a crucial role in explaining how we learn sensation-words. The point is that each of us really has his or her own internal experience that is privately known to each other. Without basing the way in which we pick out these sensations on the public framework of language, any talk of sensations becomes solipsistic. Wittgenstein's point is that this solipsistic way of speaking is private language, of which the possibility is rejected unless it is dependent on the public framework of language-games. But how does this genuinely private experience come to be dependent on the public framework?—By means of natural, primitive behavior that accompanies internal sensations. Wittgenstein in fact asks himself the following:

Now, what about the language which describes my inner experiences and which only I myself can understand? *How* do I use words to stand for my sensations?—As we ordinarily do? Then are my words for sensations tied up with my natural expressions of sensation? (*PI*, sec. 256)

Indeed, this 'tie-up' of words with natural expressions plays a crucial role by which sensations enter into language-games. If this correlation is set up, no talk of private experience are private any longer. Wittgenstein continues,

In that case my language is not a 'private' one. Someone else might understand it as well as I. (*PI*, sec. 256)

Sensations everyone has are private. But the words that express sensations cannot be private just as the facial expression of pain cannot be private. So it might really be possible that one's beetle is different from others' (*PI*, sec. 293) or that one section of mankind has sensations different from another section (*PI*, sec. 272). But it is possible for us to relate these private sensations to a publicly accessible framework. Sensation-language cannot be private. This reminds us of Wittgenstein's remarks in 1936 lecture notes:

Solipsism. //The conception of solipsism// does not stretch to games. The other can play chess as well as I.

I.e., when we play a lang[uage] game we are on the same level.[85]

In this quote, Wittgenstein, proclaiming the primacy of language-games, shows the very way in which private language loses its whole point. Everyone has his or her own private sensations. But since language is the game we play within a public framework, the sensation-words cannot be meaningful without being dependent on the public framework. Even the solipsistic way of name-giving to sensations has to be mediated by the public language-game. Indeed, this is the role of the language-game as the connecting link between language and the world. The names of sensations, in this case, too, have to be connected to sensations themselves by the language-game. Trying to give names to sensations independently of this public framework of language-games degenerates to solipsism, whose place cannot be secured in language-games.

Wittgenstein now can successfully express all the phenomenological realms of our experience in our physicalistic language. Since Wittgenstein never intends to deny the existence of what belongs to the objects of sensation-words, he has done all the work he has to do after the rejection of phenomenological language. After the rejection of phenomenological language, he has to explain how we can express immediate experience without phenomenological language. He tentatively manages to explain the way we describe immediate experience by means of the solipsistic way of expressions that is in operation as parasitic to physicalistic languages. He holds this view through his middle-period writings as he develops different ways of identifying objects. Although the idea of identification is good enough to identify any external objects as physical objects or sense-data, it does not apply to our internal experience, which is entirely private to its owner. Only I can directly experience my toothache, and the same idea applies to everyone else. In this private and solipsistic realm, the phenomenological way of language-use—e.g., phenomenological name-giving of 'this' and 'that' to each pain one has—seems a correct one. But Wittgenstein rejects phenomenological language in 1929. Hence, it is necessary for him to explain this phenomenological realm of immediate experience within the framework of physicalistic language. The answer to this difficult question is settled only when Wittgenstein develops the idea that the language-game is the most fundamental ground of our language and language-use, so that all the meaning, whether it is of the external or of the internal objects, should be given by our practice of language-games that involves public framework. So even the expression of the most private sensation, too, can be meaningful by the mediation of language-games we play. Wittgenstein does not appear to reach this conclusion until 1944 when the idea of the primacy of language-games and the private language argument begin to appear.[86]

Therefore, we can say that with the so-called private language argument, Wittgenstein finally gives all the answers he has to give in response to his rejection of phenomenological language in 1929.

CHAPTER IV

WITTGENSTEIN ON COLOR CONCEPTS

1. WITTGENSTEIN ON COLOR CONCEPTS: THE EARLY VIEW

One of the persistent themes in Wittgenstein's philosophy is the concept of color. It is evident that Wittgenstein takes up the problem of color from his early period on. He mentions color only a few times in the *Tractatus*, but with very important ideas in his phenomenology. He not only introduces the concept of color-space,[1] which consistently continues to appear in his later writings about color concepts, but treats color as one of the forms of objects.[2] The introduction of color-space that cannot be separately understood without reference to visual space suggests Wittgenstein's concern is not with physical colors but phenomenological colors from the initial stage of his philosophy.

What Wittgenstein means by phenomenological colors is not what we can know by means of physical or chemical examination, but what we know immediately from experience. His remark that color is the form of the object confirms this point. Just as we cannot perceive a stick without length, so can we not imagine a colorless object in a visual field. Color is one of the built-in logical forms in our immediate experience.[3] For instance, it is the structure of color, i.e., the logical form that determines the impossibility of a point in visual space having two different colors at the same time. For the early Wittgenstein, we know the whole conceptual structure by attending to immediate experience, because the given objects with logical forms determine the way in which the whole world is arranged. Similarly, all we have to deal with in color is the logical structure of color we would have in immediate experience. Thus, it becomes a conceptual analysis of color, and not physical, physiological, or otherwise psychological.

In this context, it is very important to see that Wittgenstein almost always has color-space in mind when he takes up color in his discussions. And, of course, it coincides with his one-time plan of building a purely phenomenological theory of color.[4] What Wittgenstein means by phenomenological color theory is "a theory in pure phenomenology in which mention is only made of what is actually perceptible and in which no hypothetical objects occur."[5] Wittgenstein's goal in this regard is to reveal the logical (or rather phenomenological) structure of colors not by adding any hypothetical elements of physics, but by the direct descriptions of what we experience immediately. This conceptual analysis is based on colors in color-space in which the geometrical representations of colors take place. Wittgenstein believes that our usual way of representing colors is misleading, and thus he keeps showing his dissatisfaction with the color-circle. The way he draws a problem from the color-circle is present in the *Philosophical Remarks* :

> At any rate, orange is a mixture of red and yellow in a sense in which yellow isn't a mixture of red and green, although yellow comes between red and green in the colour circle. (*PR*, sec. 219)

In fact, in the color-circle, orange lies between red and yellow, and yellow between red and green. But definitely we do not produce yellow by mixing red and green, as we produce orange from red and yellow. Also, while orange is a mixture of red and yellow, there does not exist a mixture of orange and violet at all.[6] For this grammatical reason, Wittgenstein sees the representation of colors by the color-circle as inappropriate, and proposes to use the color-octahedron for the right representation of color we immediately experience. Wittgenstein also thinks the color-octahedron is the better way to represent what he calls color-space, for it shows the grammatical difference immanent in different colors, which is not revealed by the color-circle that is a circular presentation of spectrum. This representation of color by the color-octahedron is not like the one that can be established empirically by experiment, and thus Wittgenstein gives it an *a priori* status:

> An octahedron with the pure colours at the corner-points e.g., provides a *rough* representation of colour-space, and this is a grammatical representation, not a psychological one. On the other hand, to say that in such and such circumstances you can see a red after-image (say) is a matter of psychology. (*This* may, or may not, be the case—the other is *a priori* ; we can establish the one by experiment but not the other.) (*PR*, sec. 1)

As the above quote shows, Wittgenstein clearly undertakes logical analysis of color when he talks about color concepts. And that, of course, has nothing to do with the empirical examination of color as physicists or psychologists might be interested in. It is no wonder, in that sense, that Wittgenstein read Goethe's *Theory of Colour*, and took it seriously, which challenges Newtonian color theory based on optical experiments.[7]

Wittgenstein is not satisfied with the color representation by means of the distribution of each color on the color-circle. He says the two statements "Orange lies between red and yellow" and "Red lies between violet and orange" involve two different grammatical usages of the word 'between'. In the first case, the expression 'lie between' means a mixture of two simple colors; in the second, it means a simple component common to two mixed colors. If this is the case, the word 'between' has two fundamentally different grammatical usages.[8] Wittgenstein's point is that when color is represented by the color-circle, we are misled to think that there is a uniform transition from color to color on the color-circle, which cannot explain the two different categorical uses of the word 'between' in the above example. In the case of primary colors, however, we do not have an image of a continuous transition, but only see the discrete hues.[9] Moreover, we cannot produce four primary colors by mixing other intermediate colors. Hence, Wittgenstein comes to say that those primary colors are points, whereas the intermediate colors are not.[10] Thus, we should not use the color-circle as the correct form of representation for color, where this difference does not show. For Wittgenstein, the better form of representation is to use a square, which explicitly distinguishes four corner-points for four primary colors, where blue, green, yellow and red are located.[11] When black and white are added, Wittgenstein naturally opts to use an octahedron, whose bottom and top corner-points would represent black and white respectively for the similar reason that the grammar of black and white is also radically different from that of the four primary colors.

To be sure, Wittgenstein's concern here is a grammatical one. What he calls grammar, as I pointed out earlier, is the virtual replacement of his early-period concept of logic, and bears phenomenological characteristic fundamentally. Color as a subject-matter in phenomenology is important in the sense that we could know its logical structure purely from what is given in immediate experience. We do not need any hypotheses or experiments. It is not what can be otherwise than is, and thus not what is a matter of taste. As the young Wittgenstein tries to read off all the possible determination of our conceptual structure from what is given in immediate experience, all the complexity of color, too, can be cleared up through the conceptual analysis of color phenomena. Wittgenstein here is concerned with grammar and thus the phenomenology of color as the following remarks suggest:

> The words 'Colour,' 'Sound,' 'Number' etc. could appear in the chapter headings of our grammar. They need not occur within the chapter but that is where their structure is given. (*PR*, sec. 3)
>
> Isn't the theory of harmony at least in part phenomenology and therefore grammar?
>
> The theory of harmony isn't a matter of taste. (*PR*, sec. 4)

For Wittgenstein, what makes harmony harmonious is not determined by human taste. Arbitrary combination of different notes at a given time will not necessarily result in a harmony of sounds. Only a certain combination of notes will do. The way in which musical notes are so combined that they produce a chord is not a matter of taste, but is somehow already built into each note. It is the very context in which Wittgenstein says the theory of harmony is in part phenomenology and thus grammar.

The reason why grammar is assimilated to phenomenology here is seen by the fact that in the use of words (e.g., in connection with color, sound, or number), certain combination of words is excluded as nonsense. Indeed, it is not conceivable to say that this color sounds harmonious, or that this color is a semitone higher than another. Now in color, Wittgenstein is concerned with a theory very similar to the theory of harmony. In the above quote, Wittgenstein could have said also that the theory of color is not a matter of taste, because he also talks about the given colors, which is the starting-point of his logical analysis.

One example that reveals the grammatical point is provided in Wittgenstein's course of argument with regard to the correct representation of color-space. It is directly related to the above-mentioned case of two different usages of 'lie between'. From a different perspective, the same grammatical point can be made as follows. We have seen that Wittgenstein makes it clear that the four primary colors can be represented as points, while intermediate colors cannot. For example, even though we have two oranges, one of which is more red-orange and the other more yellow-orange, we cannot say there is a midpoint between red and yellow. Therefore, it is nonsensical, or very misleading, indeed grammatically misleading, to apply the words 'closer to' and 'further from' to color concepts.[12] From this, it is obvious that when Wittgenstein talks about color, he undertakes logical analysis of color concepts based on color-space. His ultimate concern regarding this matter is to get at the correct representation of color-space, which necessarily involves grammatical investigations.

It is clear now that Wittgenstein's phenomenological color theory is the logical analysis of color, which at the same time is assimilated to the grammatical investigation of color in color-space. What this investigation includes becomes clear as we have seen that Wittgenstein's primary concern in the *Philosophical Remarks* is to figure out how the logical structure of the four primary colors differs from that of the intermediate colors, and also how the logical structure of the four primary colors differs from that of black and white. So we can say that from the early to the middle period, Wittgenstein's concern with color is to figure out how color terms are combined to make sense. Because colors are given with logical forms, and the logical forms determine the logic of color words, i.e., whether certain combination of color words makes sense or not, there is no doubt that Wittgenstein performs the grammatical investigation of color concepts. The specific effort made for this by him is to find the better, if not the right, representation of color-space. In fact, Wittgenstein sees that the color-octahedron, in contrast to the color-circle, is a grammatical representation of color-space.

Another decisive point which supports that Wittgenstein clearly undertook the logical analysis of color concepts is given in Wittgenstein's conversations with the members of the Vienna Circle, which proves that the position with regard to color in the *Tractatus* is also phenomenological. There Wittgenstein says:

Now let us take the statement, 'An object is not red and green at the same time'. Is all I want to say by this that I have not yet seen such an object? Obviously not. What I mean is, 'I *cannot* see such an object', 'Red and green *cannot* be in the same place'. Here I would ask, What does the word *'can'* mean here? The word 'can' is obviously a grammatical (logical) concept, not a material one.[13]

By saying that 'can' or 'cannot' in the color case means logical possibility or logical impossibility, Wittgenstein once again makes it clear that he is not concerned with empirical matter. Indeed, the matter of color-incompatibility in the *Tractatus* also has nothing to do with physical possibility or impossibility. Now, as was discussed in chapter 2, section 6 above, Wittgenstein, struggling with the color-incompatibility issue, tries to find suitable notations for color ascription that would coincide with the correct logical forms. Then, we would be able to see color-incompatibility as contradiction by the logical notation only, i.e., as logical impossibility. Of course, Wittgenstein appears to fall short of coming up with the right solution that will embrace color-incompatibility within the scheme of truth-functional logic.[14] Thus, he gives up the completeness of truth-function theory towards the middle period. But in the course of development, Wittgenstein's major issue in color from the early period is the logical possibility of color in color-space, as it is repeatedly confirmed in Waismann's records:

. . . 'A point *cannot* have two colours at the same time'. Here the word 'can' means *logical possibility* whose expression is not a proposition but a rule of syntax.[15]

For the early Wittgenstein, colors are objects with logical forms. Colors are given as all the other objects of immediate experience are given. And the logical forms of colors determine the way in which colors are combined in our experience. For example, we have color-vocabulary in our language. But the logic of this vocabulary is found in the experience of color in visual space. There are the built-in logical forms that determine how these colors are arranged. The logic of color-words, thus, is determined by the logical forms of colors. Whether a point in the visual field can have two colors at the same time is all determined by the logical forms of colors. Having built-in logical forms within themselves, colors for the early Wittgenstein are phenomenological objects. Dealing with the logic of color in immediate experience, Wittgenstein's point in color-incompatibility is indeed a phenomenological point.

Wittgenstein, in fact, points out the difference between physical impossibility and logical impossibility in the *Blue Book*:

> Let us think straight away of a similar case: "The colours green and blue can't be in the same place simultaneously." Here the picture of physical impossibility which suggests itself is, perhaps, not that of a barrier; rather we feel that the two colours are in each other's way. What is the origin of this idea?—We say three people can't sit side by side on this bench; they have no room. Now the case of the colours is not analogous to this; but it is somewhat analogous to saying: "3 x 18 inches won't go into 3 feet." This is a grammatical rule and states a logical impossibility.[16]

Wittgenstein continues to say that the bench case implies a physical impossibility that cannot usually be known without an empirical testing, and that is why the two kinds of impossibilities are confusing. He attributes the main reason for this confusion to the similar forms of expressions in our use of language. Once again, Wittgenstein makes himself clear about the direction of his discussion. Indeed, his concern is the phenomenology of color and not physics. It is the one we immediately know without any empirical testing. Hence, grammatically speaking, the sentence "The color green and blue cannot be in the same place simultaneously" is similar to the sentence "6 foot is 6 inches longer than 5 foot 6," which we immediately know, and is not similar to the sentence "He is 6 inches taller than I," which requires an empirical testing. That Wittgenstein's logical, or rather grammatical, point in color does not require any empirical testing is directly related to his middle period attitude that his theory of color is about the color we recognize immediately and absolutely free of hypotheses.

Thus, it is natural to see that Wittgenstein assimilates grammar to phenomenology. He finds that logic lying behind immediate experience should be found by means of the grammatical investigation of language we use in attributing what is given in immediate experience. Even after the rejection of phenomenological language, Wittgenstein's phenomenological analysis is possible, for our language still mirrors reality. The way reality appears to us is grammatically (i.e., logically) determined. Indeed, this grammatical investigation amounts to finding out the correct representation of what is given, through the recognition of what is essential and what inessential in the use of our language. This grammatical point exactly coincides with the earlier example of the two different grammatical usages of 'lie between' in the color case. Through the grammatical analysis of 'lie between' in the color case, Wittgenstein reaches the conclusion that the color-octahedron should be the better form of representation for color-space than the color-circle. And this is nothing other than Wittgenstein's attempt to achieve the same goal as phenomenological language would have done. Contending that phenomenological language is no longer possible, Wittgenstein tries to have the same job done by the grammatical investigation of what is given in immediate experience.

2. THE LANGUAGE-GAME OF COLOR

From the late middle period, Wittgenstein's treatment of color concepts begins to be coupled with his newly developed idea of language-games. In the early middle-period writings such as the *Philosophical Remarks*, after the rejection of phenomenological language, Wittgenstein still tries to find a suitable way in which color-words represent immediate experience of color. However, the writings from the *Brown Book* to the *Philosophical Investigations* and the *Remarks on the Philosophy of Psychology* indeed hardly contains what he calls phenomenology or phenomenological accounts of color. With the introduction of the language-game idea, Wittgenstein's accounts of phenomenology begins to disappear, as he believes the analysis of grammar within physicalistic languages would do the same job that phenomenological language would have done. However, this should not be understood as if Wittgenstein had given up the phenomenological attitude toward color along with phenomenological language, for there is hard evidence in his very last writings entitled *Remarks on Colour* which assures us that Wittgenstein in fact has never stopped treating color from the phenomenological viewpoint. Without phenomenological language that describes color in color-space, the first thing Wittgenstein has to do is to secure the way in which we talk about color. Needless to say in this regard, Wittgenstein's way of doing this is to present the language-games of color, which amounts to the same job the phenomenological language of color would have done.

Wittgenstein's discussion of the language-game with color-words appears as early as in the *Brown Book*, where he presents the idea of language-games as the complete system of human communication.[17] Explaining what the language-game of color consists of, Wittgenstein introduces a thought-experiment in which one orders another to bring a piece of cloth of a certain color-sample.[18] Wittgenstein's attention is drawn to the way in which color-words are used by showing what is involved in the acts of the players of this game. In particular, Wittgenstein pays special attention to how one behaves in choosing the right piece of cloth that matches the color-sample shown to him. The obvious reason for Wittgenstein to bring in this thought-experiment is to show that the language-game of color depends on the public framework as any other language-game does. It is an attempt to answer the question of how color-words are used, or how the meaning of color-words is brought about. So the thought-experiment goes on by examining the possible behavior of the one who is supposed to choose and bring the right piece of colored cloth. One possible case is to rely on color-charts as a public framework, so that one brings a color-chart with him and makes a color-comparison for the matching colored cloth. But this possibility is simply ruled out because we can easily see that it is not the usual way of using color-words in everyday life. Nobody carries his or her own color-samples or color-charts in everyday life. Of course, this does not mean that we never use color-charts or color-samples in our use of color-words. They might in fact contribute to identification or our use of color-words. The important thing is that color-samples cannot be used as the sweeping argument for the source of meaning in color-words.

Then, another possibility for choosing the right color is to rely on the memory image of color in one's mind's eye. As Wittgenstein once attempted to construct the conception of time relying on our memory of earlier and later, memory can serve as a basis of our phenomenological discourse.[19] But Wittgenstein does not give conceptual primacy to our memory as the fundamental basis of his color-discourse. The point is obviously shown in his other late middle-period writings "Notes for the Lecture on 'Private Experience' and 'Sense Data'." There he asks himself,

Is it ever true that when I call a color 'red' I serve myself of memory?? /make use of memory?/[20]

Again, in this case, Wittgenstein does not mean to deny that we in fact have a memory image of color when we identify color and use color-words. Rather, the point he makes with this remark is a conceptual one implying that the meaning of color-words does not come from the memory image we might have from the past color-experience, because the introspective means such as memory can never serve as a public framework required in language-games. There is a language-game that is being played, and color-charts or memory images of colors or any other color-experiences work as constitutive parts of the language-game of color. So Wittgenstein gives us warning of the misuse of memory image.

To use the memory of what happened when we were taught language is all right as long as we don't think that this memory teaches us something essentially private.[21]

This is indeed an enlightening remark if we consider Wittgenstein's ultimate position that language-games are the most fundamental ground of semantics, which has been discussed in the previous chapter. There, it was argued that even the most private experience, when expressed in language, cannot rely on the introspective means like memory or any other rules for its meaning, but should be mediated by language-games. So the quoted remark, by all means, is Wittgenstein's point against the view that rules can give the meaning of words, because either relying on color-samples or memory images is nothing other than an attempt to describe rules of the language-game, implying that rules can give the meaning of color-words. But essentially, this attempt has to fail, because Wittgenstein's view is that memory can serve as only one constitutive part of the language-game of color.

Needless to say, memory cannot serve as a viable criterion for the identification of color, for sometimes we experience 'darkening of memory'.

> But can we not sometimes speak of a darkening (for example) of our memory-image? Aren't we as much at the mercy of memory as of a sample? (For someone might feel like saying: 'If we had no memory we should be at the mercy of sample.')—Or perhaps of some chemical reaction. Imagine that you were supposed to paint a particular colour 'C', which was the colour that appeared when the chemical substance X and Y combined.—Suppose that the colour struck you as brighter on one day than on another; would you not sometimes say: 'I must be wrong, the colour is certainly the same as yesterday'? This shews that we do not always resort to what memory tells us as the verdict of the highest court of appeal. (*PI*, sec. 56.)[22]

We cannot rely on our memory image of color when we identify and use the word for color. It lacks a public framework, as it does in the case of private sensations, and thus can give no meaning to the color-words. Having a memory image of color is, as we sometimes can make use of color-samples for color-comparison, one useful tool that constitutes the whole texture of the language-game played with color-words. We cannot really pin down any specific rules that contribute to the meaning of color-words, because there are no such rules or specific experiences that would explain how the meaning of color-words is brought about. In fact, any approach based on one or two specific features of the way that color-words are used will result in an unsuccessful account of the language-game of color. Wittgenstein's suggestive account in the *Brown Book* indeed casts light on this point:

We find that what connects all the cases of comparing is a vast number of overlapping similarities, and as soon as we see this, we feel no longer compelled to say that there must be some one feature common to them all. What ties the ship to the wharf is a rope, and the rope consists of fibers, but it does not get its strength from any fiber which runs through it from one end to the other, but from the fact that there is a vast number of fibers overlapping.[23]

To see that there is a vast number of fibers overlapping in a rope causes us not to rely on any specific feature in the use of color-words. Rules cannot serve as the ultimate source of meaning. Rather, what we have to see more importantly is the fact that there is the language-game of color which is being played. There can be no specific reason that one sees a red thing and calls it red. One only plays the language-game of color in the way in which one has been taught to. Therefore, the way we use certain color-words, or the way we play the language-game of color needs no justification. It is our activity of playing the language-game of color that justifies the meaning of this or that color-word. Concentrating on memory image or any other specific feature in our use of color-words is similar to the case where one tries to explain the strength of a rope by the account of each individual fiber ignoring the fact that all the fibers are interwoven to form a rope. This point is much more clearer in the later part of the *Brown Book*. There Wittgenstein gives an illuminating example:

Now to the question "What made you use the word 'darker'. . .?" the answer may be "Nothing made me use the word 'darker',—that is, if you ask me for a *reason* why I use it. I just use it, and what is more, I used it with the same intonation of voice, and perhaps with the same facial expression and gesture, which I should in certain cases be inclined to use when applying the word to colours."[24]

What decides the meaning of color-words cannot be found anywhere else than in the language-game. Indeed, what the language-game decides is the verdict of the highest court of appeal, and this is not an exception for the case of color-words.

In this sense, it is no surprise at all that Wittgenstein directs his attention toward the teaching and learning of color-vocabulary in

Remarks on the Philosophy of Psychology.[25] That one can say "This color is red" means one knows how the color-word is used and one grasps the technique to use the color-word. What an adult teaches a child in this case is an ability to behave in such a way that the child can bring something red when an order is given. Also, teaching and learning of the language-game of color lead people to behave in such a way that they do not conflict with each other in describing the same thing as red or green. Thus, the use of color-words, too, should be referred to the public framework of language-games to be meaningful. That, of course, is not relying on any specific feature of our language-use such as color-charts or color-samples, but on the language-game itself.

One interesting thing regarding Wittgenstein's discussion of the language-game of color is that he consistently treats it as a case similar to the language-game of sensations. As shown in an earlier quote that criticizes the memory image of color, Wittgenstein's approach reminds us of his discussion of pain-words. His criticism of relying on the memory image of color is indeed parallel to his criticism of the ostensive way of name-giving to private sensations in the *Philosophical Investigations*. While discussing the color case, Wittgenstein in fact illustrates an example that enables one to see an interesting similarity between the way in which the words for pain and color operate. An example similar to that of diary-keeper, which appears in section 258 of the *Philosophical Investigations*, also appears in his 1936 lecture notes. But in this case, what the diary-keeper takes notes of is not private sensations but the occurrence of a particular color-sensation. Wittgenstein says,

> But I could use language just for making entries in my diary and without ever having learned it. I could have invented a name for the particular colour sensation, say, the name 'red' and then used this name to note down whenever I had that colour sensation. That means, you (would) play a private language game with yourself. But let's see, how are we to describe this game?—Christening. The words "'seeing red' means a part[icular] experience" are senseless unless we can follow them up by namely this → (pointing) or else they may say experience as opposed to phy[sycal] ob[ject], but then this is grammar.[26]

The diary-keeper who has never learned the name of a particular color-word would try to establish the name of the color he experiences by taking notes when the same color-experience occurs. He may rely on the introspective means such as memory image to re-identify the particular color. But this, lacking in a public framework, will result in failure of establishing any connection between the expression of color and the experience of color. The name of color can only have meaning by reference to the language-game. Discussion of rules apart from language-games cannot explain the way in which meaning is brought about, "for what we call the meaning of the word lies in the game we play with it."[27] Certainly, the diary-keeper of color in this example develops into the diary-keeper of private sensation in section 258 of the *Philosophical Investigations* as follows:

> Let us imagine the following case. I want to keep a diary about the recurrence of a certain sensation. To this end I associate it with the sign 'S' and write this sign in a calendar for every day on which I have the sensation.—I will remark first of all that a definition of the sign cannot be formulated.—But still I can give myself a kind of ostensive definition.—How? Can I point to the sensation? Not in the ordinary sense. But I speak, or write the sign down, and at the same time I concentrate my attention on the sensation—and so, as it were, point to it inwardly.—But what is this ceremony for? for that is all it seems to be! A definition surely serves to establish the meaning of a sign.—Well, that is done precisely by the concentrating of my attention; for in this way I impress on myself the connexion between the sign and the sensation.—But "I impress it on myself" can only mean: this process brings it about that I remember the connexion *right* in the future. But in the present case I have no criterion of correctness. One would like to say: whatever is going to seem right to me is right. And that only means that here we can't talk about 'right'. (*PI*, sec. 258)

Wittgenstein's objection against the diary-keeper's ceremony is that in neither case the public framework is involved. Memory cannot serve as the proper means by which color-sensations or private sensations can be publicly referred. On the ground that language-games determine meaning and rules, we can understand the above case. Nonetheless, what puzzles us is Wittgenstein's attitude toward the subject-matter of private sensations and color sensations which apparently are different in character in the sense that colors are outwardly identified as we see, whereas private sensations such as pain can never be outwardly identified. Wittgenstein's assimilation of the color case to private sensations seems more than serious as witnessed by the following remarks:

I treat colour concepts like the concepts of sensations.[28]

The colour concepts are to be treated like the concepts of sensations.[29]

Then, the question is why Wittgenstein treats color-words as analogous to sensation-words. Regardless of the apparent difference in the two cases, we can figure out the very similar characteristic, which is believed to be Wittgenstein's point, of the ways in which color-words and sensation-words operate.

As was discussed in the previous chapter, Wittgenstein makes it clear that our pain-words replace some of the primitive pain-behavior such as crying, exclaiming or grimacing. But it was pointed out that that does not mean the meaning of a pain-word is pain-behavior. Indeed, it is absurd to say that "I have a toothache" means crying, grimacing or any other natural expressions of pain. On the contrary, a pain-word means pain. 'Toothache' is the name of pain in a tooth operated in the language-game of pain. And in a broader perspective, the pain-word, too, belongs to pain-behavior. A pain-word can be as spontaneous as pain-behavior. Therefore, no other elements can come in-between the pain-word and the pain itself. We cannot drive a wedge between pain-words and pain. Pain-behavior might directly reflect the internal and private state of pain inside one's mind, and a pain-word too is a spontaneous reaction to pain-experience, only in this case mediated by the practice of a language-game we learn to express different pains in different cases. Our pain-experiences are private, but pain-words cannot be private. No matter how private a pain may be, it cannot be expressed in words without reference to physicalistic language of pain in the practice of the language-game. Language-games directly link pain-words and pain. There can be no other mediator between them.

Similarly, the way we use color-words is equally spontaneous as the way we use pain-words. We cannot really drive a wedge between the color-impressions we get and the color-words. Everyone has his or her own color-impressions. But they cannot be expressed in words without depending upon the physical framework of language. The color-impressions we get are direct experience, and the color words are directly linked to the color-impressions by means of language-games. Color is also the case where language and reality have a direct interface. As the language-game of sensations is a primary language-game, so is the language-game of color a primary one. No matter how phenomenological they may be, there is no other way for us to refer to color-impressions than in our physicalistic color-words operated in language-games. Neither in the case of the way color-words are used, nor the case of the way pain-words are used, do we put a label on our perception.

Wittgenstein's points with regard to color concepts and the way they operate more or less have equivalent counterpart in the evidence of cognitive science. According to research in cognitive science, there is evidence which supports the idea that color concepts do not operate in a purely perceptual way. Researchers say there is a case where a patient is unable to name colors or point to them while there is no disturbance in color-sensations.[30] Thus, regardless of perceiving color-impressions correctly, say red, the patient is not able to identify and name the color he sees as red. This is exactly the opposite case of color-blindness, and indirectly proves Wittgenstein's point that color concepts do not operate through the ceremony of ostensive labeling on unedited raw data of color-impressions. Of course, this does not mean that research scientists have empirically proved what Wittgenstein said. Instead, what is relevant here is Wittgenstein's insight with regard to our color-experience and the way we use color-words to express it—i.e., having the color-impressions of red is not enough for us to be able to use the word 'red'. The color 'red' we immediately experience is one of Wittgenstein's concerns, and it is obviously what he means by "the world we live in is the world of sense-data."[31] Unfortunately, however, Wittgenstein no longer has phenomenological language that faithfully describes the pure impressions of red. The words do not operate in such a way that we label on a thing. To be able to express the impressions of red, we need to refer to the physicalistic language system, which is implied in his words "the world we talk about is the world of physical objects."[32] Wittgenstein's ultimate solution, of course, is to rely on the language-game of color we play. But his thought-experiment that supposes a person who carries a color-sample along in order to identify the correct color vividly shows Wittgenstein's insight that we need more than passive reception of data. And this thought-experiment is worth being appreciated in light of today's developments in cognitive science.

One important example with regard to the analogous approach of the way color-words are used to the way sensation-words are used is present as interrupting remarks on color in the middle of the so-called private language argument. Apparently, sections 273-280 of the *Philosophical Investigations* are unexpected insertions between the discussion of how the meaning of sensation-words is brought about. At bottom, however, it has to be understood as Wittgenstein's careful arrangement that signals how sensation-words are analogous to color-words.[33] There, Wittgenstein focuses on the peculiar characteristics of color-words that tend to differentiate the color-impression which a perceiver has and the physical color known to everyone.[34] It is, though, a misleading case, viewed exclusively from the perceiver's own private experience of color-impressions, as in the case that one tries to set up the meaning of color-words from one's memory of color. But that is what happens when one thinks as if one can detach the color-impression from the physical color.[35] That of course is not possible for Wittgenstein. It is wrong in the sense that one believes one can refer to one's color-impression independently of a publicly available framework of communication. It is an attempt to make a distinction between one's color-impression and the physical color, and is roughly analogous to the case in which one tries to establish private use of a word based on one's private experience only, or in which one claims that "another person cannot have my pain." What goes wrong in this case of private experience or the pain case is that it presupposes one can express such things as pain in a purely phenomenological way without being mediated by language-games. It is the case where one thinks one's pain is real only for oneself. Similarly, one is tempted to think as if one could express color-impressions without reference to physicalistic vocabulary of color. But that is what Wittgenstein points out as a temptation. We cannot make a distinction between color-impressions and the physical color, just as we cannot do so between pain and pain-word.[36] What we experience are indeed phenomenological data, but those cannot be expressed in any other way than with physicalistic language.[37]

Thus, we now understand the point made by Wittgenstein's assimilation of the color case to the sensation case. In a similar manner that he does not reject the existence and role of 'beetle' in the sensation-words, so does he not deny the fact that we have memory image of color or color-impressions that constitute the language-game of color. All he rejects in the color case is that we can meaningfully speak of color without reference to physicalistic color-words operated in the language-game. So it is, in a way, a very logical step for Wittgenstein after his rejection of phenomenological language. The only language we have is physicalistic language, and that is found in everyday practice of language-games. Color-words, too, have to be grounded on this rule-governed human activity of language-games. The phenomenological character of color does not go away, but still Wittgenstein is successful in explaining how this phenomenological color is expressed in physicalistic language.

3. WITTGENSTEIN ON COLOR CONCEPTS: THE LATER VIEW

Wittgenstein's very last writings, written in the last days of his life, have been published as *Remarks on Colour*, along with *On Certainty*, and prove what occupied Wittgenstein's mind through the last days of his life were indeed phenomenological problems. What he purports to do in his discourse on color is clearly stated in the following remarks:

> We do not want to establish a theory of colour (neither a physiological one nor a psychological one), but rather the logic of colour concepts.[38]

This vividly shows that the original concern Wittgenstein had in the earlier days of his life has not changed through his very last days. Wittgenstein's words 'the logic of colour concepts' cannot mean anything other than phenomenology of color as shown earlier. This phenomenological analysis of color concepts has nothing to do with physics in the sense that it can "neither agree with nor contradict physics."[39] Not only that, Wittgenstein is not concerned with Gestalt psychology either, for his question in the color-discourse is "what is the meaning of this expression, or what is the logic of this concept?"[40] In the middle-period writings, this question was presented as the right representation of color in color-space. In particular, Wittgenstein's attention was primarily directed toward the four primary colors that differ from intermediary colors in logic. Thus, his discussion there includes that the sentence "Red lies between yellow and blue" is grammatically different from the sentence "Orange lies between yellow and red." This logical or grammatical point, according to Wittgenstein, could be represented by the color-octahedron, which says that "you can speak of a reddish blue but not of a reddish green, etc."[41]

Now the old Wittgenstein takes up this problem and again asks why we cannot have expressions such as 'reddish green' or 'yellowish blue'.[42] The basic approach to this matter has not changed in the sense that Wittgenstein still sees the problem in connection with the logic (i.e., grammar) of color concepts, or as he sometimes prefers to say, color-geometry. This time, however, he relates this logical point with the idea of the language-game. So he pays attention to how color-words operate to make sense of some expressions and to exclude others as nonsense. The way this conceptual point is related to the language-game is shown in the following example.

> If I had taught someone to use the names of the six primary colours, and the suffix "ish" then I could give him orders such as "Paint a greenish white here!"—But now I say to him "Paint a reddish green!" I observe his reaction. Maybe he will mix green and red and not be satisfied with the result; finally he may say "There's no such thing as a reddish green."—Analogously I could have gotten him to tell me: "There's no such thing as a regular biangle!", or "There's no such thing as the square root of -25."[43]

Whether we could have expressions like 'reddish green' relies on logical possibility as the possibility of regular biangle is excluded by logic. Now, Wittgenstein holds that the logical difference lying behind the various color-words is determined by the language-game. What makes green a primary color and not an intermediary color between yellow and blue is decided by the language-game we play.[44] Likewise, what excludes 'reddish green' as nonsense, too, is decided by the language-game. The way color appears to us cannot be grasped without a conceptual framework. When he believed in phenomenological language, Wittgenstein thought that immediate representation for immediate experience of color was possible. But Wittgenstein realized that the phenomenological world we live in cannot be expressed without being mediated by means of physicalistic language. The very logical (i.e., phenomenological) point cannot be discussed without a connection to the language-game. So his middle-period phenomenological problems are replaced by the grammatical problems in language-games.

This new perspective brought in with the supreme importance of the language-game, however, gives Wittgenstein more complicated problems with regard to his treatment of color concepts. For example, he extends his concern from what we can call surface colors to what requires many different concepts as colors appear in many different language-games. Wittgenstein's initial concern might have started from the different logical structures between primary colors and intermediary colors, and also from the unique characteristic of black and white. His middle-period solution is to represent color-space with the color-octahedron which shows the grammatical difference between the colors in question. But Wittgenstein seems to realize that there are other dimensions to color-space than what we simply can represent with the color-octahedron.

For example, in everyday life, we not only see objects with colors on their surfaces such as books, tables, and chairs, but also objects of which colors do not appear to lie on their surfaces such as colored glass or colored liquid. In order to explain this variety of color phenomena Wittgenstein brings the concepts of transparent and opaque colors into his discussion which require the dimension of spatial depth in visual space. A typical example is to ask "Why is it that something can be transparent green but not transparent white?"[45] Also, Wittgenstein brings in the contrast between luminosity and cloudiness of colors.[46] To be sure, he finds a solution to the question why we cannot have a white transparent object in the logic of color concepts. That grey is a cloudy color and we cannot conceive of something as being 'glowing grey' indeed belongs neither to the physics nor to the psychology of color.[47] To understand how we combine color-words and make sense of them, i.e., to understand the logic of color-words, we have to see how the language-game of color operates.

This is particularly interesting with regard to the role of the language-game in determining a logical or conceptual point, because Wittgenstein makes clear the fact that we have transparent and opaque colors and not just surface colors does not mean that "we would use different greens to paint a piece of green glass and a green cloth in a picture."[48] The point is that green can appear to us as surface color, as in the case of green table at one time, and as transparent color, as that of green glass at another. On the other hand, white cannot be presented to us as white transparent glass, while it can be an opaque color as milk cannot be clear. In either case, Wittgenstein does not mean that we have different color-impressions, or the impressions of green or white come about differently under such and such conditions. It is not one of his concerns whether color-impressions are presented to us according to physical conditions, because Wittgenstein never intends to investigate causal connections.[49] On the contrary, his concern is the impression of a certain context. It is about how a specific green color stands in our conceptual network. Thus, Wittgenstein really makes a logical point in this complexity of color-phenomena and never intends to do what Gestalt psychologists would do. Wittgenstein says:

> Psychology connects what is experienced with something physical, but we connect what is experienced with what is experienced.[50]

We connect what is experienced with what is experienced so as to form color concepts. And this is done by the practice of the language-game of color. The logic of color concepts thus is determined by the way we practice the language-game, and also is revealed through the grammatical analysis of this language-game.

For Wittgenstein, what we have in our experience is important. But what is experienced alone is not sufficient for us to discuss our experience. For instance, we might have color-experience. But that is not the whole story in having color concepts. Logic that governs color concepts will not come about solely from our experience of color-impressions. The conceptual framework is determined by the practice of language-games. Wittgenstein, in fact, illustrates some examples that come very close to this idea, although he never expresses it explicitly. One example is the case of a black-and-white photograph (or film) where all the colors are presented by shades of black, white, and grey. What Wittgenstein questions here is why we can recognize someone's hair as blond, even though the color-impressions we have from the photograph is grey.

I see in a photograph (not a colour photograph) a man with dark hair and a boy with slicked-back blond hair standing in front of a kind of lathe, which is made in part of castings painted black, and in part of smooth axles, gears, etc., and next to it a grating made of light galvanized wire. I see the finished iron surfaces as iron-coloured, the boy's hair as blond, the grating as zinc-coloured, despite the fact that everything is depicted in lighter and darker tones of the photographic paper.[51]

What we really see in the photograph cannot be the impression of blondness in the case of boy's hair in the above example. But still we are able to see it as blond hair. Or perhaps, we should say that we are able to interpret it as being blond hair. When we see a shade of grey in the photograph and say 'blond hair', we not only have grey color-impressions, but also connect those particular impressions with earlier impressions we might have had in the actual acquaintance with blond hair. This indeed is not a simple matter to be analyzed easily. But still, what Wittgenstein intends to say seems clear with regard to our ability to form color concepts. In other words, color concepts cannot come about without the practice of language-games. That we simply have color-impressions alone in a purely perceptual manner will not be sufficient for us to be able to form color concepts in this sense. What is more important is how these impressions stand in the entire context of a conceptual network. Without doubt, Wittgenstein's answer to this question is that the way color concepts operate is determined by the language-game, which is basically a physical framework. In this context, Wittgenstein's following remarks are especially instructive with regard to our color-experience and the way in which we use color-words.

In everyday life we are virtually surrounded by impure colours. All the more remarkable that we have formed a concept of *pure* colours.[52]

Lichtenberg says that very few people have ever seen pure white. Do most people use the word wrong, then? And how did *he* learn the correct use?—On the contrary: he constructed an ideal use from the actual one. The way we construct a geometry.[53]

This point is related to the case where we only see a black-and-white photograph and still be able to recognize the blond color of a boy's hair in the actual world. For Wittgenstein, the reason we can form the concept of pure white, regardless of the fact that most of us only see lighter or darker shades of it, is the color-impression we have can only be conceptualized in context of the other concepts we have acquired through the practice of language-games of color.

There is one example in phenomenological psychology which makes a similar point to Wittgenstein's treatment of color concepts. David Katz, in his book, *The World of Colour*, explores the psychology of color from the phenomenological viewpoint.[54] Regardless of the fact that Wittgenstein's concern is not the same as that of psychologists' account of color, what makes Katz's discussion interesting with regard to Wittgenstein's view on color is that Katz also entertains various color concepts such as film colors, surface colors, volume colors, etc. Moreover, at some point, what Katz states sounds similar to what Wittgenstein says. For example, Katz says, "Our interest in colour is not the interest of the physicist, nor again is it with those aspects of colour which puzzle the physiologist. Our concern is rather with the purely psychological problems of colour."[55] One of the reasons that Katz excludes the physicist's interest in color is because he believes that physicist's approach to color is made under the highly artificial conditions of the laboratory. He maintains that most people never see the pure spectral color in their lives.[56] Rather, the way most people perceive color depends on the rest of what they perceive such as the concepts of space, surface, field, liquid, etc. The significant thing in this regard is that Katz uses the concepts of space, surface, field and liquid in classifying color concepts. What this implies is, while we might be able to have pure impressions of color, the way we recognize and conceptualize color is closely related to the ascriptions of physical objects such as space, surface, field, or liquid. There would be no complete explanation of the way in which we perceive color-impressions without relating the color-perception to other related concepts that designate physical objects. Even though what Katz does is the psychology of color-phenomena, the underlying spirit is very close to Wittgenstein's treatment of color concepts in the sense that, for Wittgenstein, language-games that give color-words meaning depend partly on the way we manipulate physical things. As was discussed in the previous section, we cannot make a distinction between the color-impressions and the physical color; the language-game is basically a physical framework, and even purely phenomenological color-impressions cannot be expressed without being mediated by the physical framework. One example that explains this point can be found in Wittgenstein's mention of transparent color. What Wittgenstein means by transparent color is something like the color of varnish. When we apply varnish to a stool made of wood, it gives us a certain color which illuminates the surface of wood, although it is transparent. Now, Wittgenstein's point is that we have the use of the word for transparent color in connection to the physical things like the surface of wood, and that the actual manipulation of physical things contributes to the way in which the language-game of color-words work.

That Wittgenstein frequently conducts the thought-experiments about color-blindness and blindness is very suggestive in this regard, for color-blindness or blindness are cases where the sources of color-impressions are either partly or entirely shut off by some causes. Wittgenstein in fact tries to see what would happen in the case where there are no color-impressions available. In the black-and-white photograph case, Wittgenstein shows how the normal person would judge the color of a real object when shown the black-and-white picture of it. Without a doubt, those with normal vision will have little difficulty in communicating about what color he or she may see.

Now, the question is asked in the case of the color-blind and the blind.

A person who is red-green color-blind would not be able to distinguish red from green as a person with normal vision does. But still, the color-blind person will use the words red and green, even though the way he or she uses them will definitely be different from that of the person with normal vision. The problem is that the color-blind person will not be able to play the language-game of color in the same way in which the person with normal vision does. The color-blind person might have the concept of color-blindness. But he or she would not have the same concept of color-blindness as the person with normal vision does.[57] The reason for this is that the way the color-blind person establishes the concept of color-blindness should be different from the way the person with normal vision does.

We may have a similar result in the case of blindness. Although a person with normal vision might be able to teach a blind person the difference between having a sight and being blind, the concept of blindness the normal person has will be different from that of the blind person. Indeed, if the normal person tells the blind person of primary colors, it will be meaningless to the blind person because he or she is not able to play language-games with color-words. Some language-games the normal person can learn cannot be taught to the blind person. And in this case, the concepts the normal person can have might easily be meaningless to the blind person. Hence, the concepts of words the blind person has would be different from those the normal person has. This confirms Wittgenstein view that the color-impressions we have, or the difference of color-impressions between the normal person and the color-blind person, is reflected in language-games we play. Not only that we have color-impressions, but that we play language-games in which those impressions are reflected is also important in the way in which we have this and that logic of color concepts. It is about the way colors appear to us. And it might be thought that color is a purely perceptual phenomenon. But Wittgenstein holds that it cannot be understood without being mediated by language-games, because language-games determine the way in which a conceptual network is set up. And the way we have this and that color concepts and not others (e.g., we have yellowish red, but not reddish green, or we have transparent green, but not transparent white, etc.) is determined logically in this regard. Thus, when Wittgenstein says the lives of the blind are different from those of the sighted,[58] they are indeed logically (i.e., phenomenologically) different.

Color concepts we have are determined by the way we play the language-game of color in which the color-impressions are reflected. And this language-game will decide which expressions make sense and which do not. So the fact that we have color-impressions alone will not explain the way we use color concepts. This point has indeed been entertained by Wittgenstein in the writings that precede *Remarks on Colour*. As was mentioned in the previous section, the thought-experiment in which color concepts are operated by color-charts, i.e. the case where color-charts are necessary in identifying and naming each color, are the case where Wittgenstein purports to show that color-words do not merely operate perceptually.[59] If color-words operate in a purely perceptual way, we could name a certain color-impression as we label an object. But color-words do not operate in such a way that we can label a thing. And this is the exact point Wittgenstein makes in section 49 of the *Philosophical Investigations*.[60] Color concepts, as Wittgenstein shows through his thought-experiment, do not operate as if we simply have color-impressions alone. As in the case of sensation-words, for Wittgenstein, the object and designation model (i.e., the '*Bezeichnung*' model)[61] will not work as a proper semantical model. We not only need to have color-impressions but the actual practice of language-games where those impressions can be reflected to form color concepts.

This remarkable point that our conceptual power is determined not by perceptual impressions alone but by means of language-games confirms again that the very phenomenological realm of our experience can only be meaningful by means of the physical framework of language-games. And this is indeed the way we connect what is experienced with what is experienced so as to form our color concepts. The logic of appearance of color is phenomenological, but it cannot be expressed in a purely phenomenological way. The language-game we play is physicalistic and its grammar can only be expressed in physicalistic language. Perhaps Wittgenstein's rather cryptic remarks in *Remarks on Colour* suggests the very problem of this aspect:

> There is no such thing as phenomenology, but there are indeed phenomenological problems.[62]

> There is indeed no such thing as phenomenology, but there *are* phenomenological problems.[63]

CHAPTER V

WITTGENSTEIN ON ASPECTS

1. WITTGENSTEIN'S USE OF THE WORD 'ASPECT'

Wittgenstein's early philosophy can be characterized as phenomenology in the sense that his aim was to describe the world as it is, without any hypotheses, just as phenomenological physicists like Mach. But for him the world ("the world we live in") is a phenomenological world. Hence hypothesis-free description and phenomenological description became for Wittgenstein to mean the same thing. According to Wittgenstein, the phenomenological world consists of phenomenological objects that can be picked out by the words 'this' and 'that' with pointing gestures, which would give immediate description of immediate experience. As his middle-period writings show, the words 'this' and 'that' are without hypothesis and thus pure and primary. But he notices that the logical behavior of these words, along with 'here', 'now', and 'I', easily prompt solipsistic interpretations. So after the rejection of phenomenological language, he tries to examine how the logic of these words is related to solipsism and how we can prevent any expressions from becoming solipsistic. As long as solipsism is merely construed as a special notation within the overall physicalistic notations, it can be rendered harmless. But if it is thought of as leading to purely phenomenological language, Wittgenstein has to disarm it. His discussion of solipsism is thus only one aspect of his defense of his new vision of language and its relation to the world.

No matter what analysis he undertakes regarding the matter, however, the words 'this' and 'that' are the most obvious examples from the Tractarian period that depict phenomenological objects faithfully. For example, when I open my eyes, I find a certain figure in red in my visual field. The most immediate description of this experience would be "I see this," or probably just "This," for the words "I see" in the sentence "I see this" may be redundant to the person who utters the sentence. If I am the speaker of the sentence, indeed "I don't wish to tell myself that it is *I* who see this, nor that I *see* it."[1] Thus, in the strictest sense, the most immediate (and thus phenomenological) description of what I see must be just saying "This," with possible pointing gesture.

Wittgenstein's Tractarian language scheme is indeed shaped the way just described. As he realizes, however, the immediate description of what is given in my immediate experience is prone to solipsism, Wittgenstein rejects phenomenological language in favor of physicalistic language. A language that relies on purely phenomenological description of the given leads to solipsism in the sense that it cannot account for interpersonal communication. It also leads to the solipsism of the present moment in that the only things that are directly given are the objects of my present experiences.

As Wittgenstein often calls them, the expressions like "I see this" or "This!" are primitive and primary language that speaks for immediate experience. But the middle-period Wittgenstein warns us that this kind of expression would be solipsistic and thus often meaningless. When I say "I see this," it makes perfect sense, and is always true to me. However, what the word 'this' may describe can well be obscure or unclear to others unless additional information is available. He concludes that the exclusive use of immediate description is not the way our language works. He thus suggests that we usually say "I see a red circle" instead of "I see this." Using 'a red circle' instead of 'this' is indication of the underlying scheme of Wittgenstein's philosophy. It indicates, basically, a change from solipsistic to non-solipsistic expressions. It means that we have to use the expressions in such a way that they include physical attributes. Most broadly, however, it means that we have to resort to the public mode of identification instead of the perspectival mode. Indeed, 'a red circle' refers to an entity that can be publicly identified, whereas what 'this' denotes need not be.

What is especially instructive in this context is Wittgenstein's use of the word 'aspect', or more commonly its German cognate

'*Aspekt*'. He frequently uses the word in his writings from 1947 to 1949, which by now have been published as *Last Writings*, *Remarks on the Philosophy of Psychology*, and the second part of *Philosophical Investigations*. Wittgenstein in these materials uses the word 'aspect' (*Aspekt*) along with aspect-seeing or aspect-change, so that readers are easily misled to think his primary concern in using the word is something like Gestalt psychology or philosophy of psychology *per se*. It is true that Wittgenstein discusses on various psychological concepts, and examines Köhler's ideas. However, his use of the word 'aspect' in these late writings is essentially the same as the way in which the middle-period Wittgenstein uses the word to criticize his earlier mistakes in the *Tractatus*. At one point in the *Brown Book*, he says,

> The danger of delusion which we are in becomes most clear if we propose to ourselves to give the aspects 'this' and 'that' names, say A and B.[2]

Wittgenstein here criticizes his earlier view that the relation between name and object are set up directly, and thus names like 'this' stand for what is given in immediate experience. Then, Wittgenstein's 'aspect' here does not have its familiar meaning of 'side' or 'facet', but rather it means a way of appearing.[3] For Wittgenstein, the word 'aspect' refers to a phenomenological object picked out by 'this' and 'that'.[4]

That Wittgenstein's aspect means the way an object appears to us is well supported by his own use of the word in other middle-period writings. In particular, the *Philosophical Grammar* gives us a few examples that really show Wittgenstein's '*Aspekt*' means the way an object appears to us. The context of the passage I quote below may not be entirely relevant to the general point of my discussion, but still it serves to be an excellent example of Wittgenstein's use of the word '*Aspekt*'.

> Someone meets me in the street and my eyes are drawn to his face; perhaps I ask myself "who is that?"; *suddenly the face begins to look different in a particular way*, "it becomes familiar to me"; I smile, go up to him and greet him by name. . .[5]

The italicized portion of the above quote is a translation of the original German; "*plötzlich ändert sich der Aspekt des Gesichts in bestimmter Weise, 'es wird mir bekannt'*." What is important here is that the English translator of the *Philosophical Grammar* did not use the English word 'aspect' for '*Aspekt*', and instead translated the phrase in question in such a way that it implies "suddenly the face appears to me differently in a particular way." From this translation, we can easily see that Wittgenstein's use of '*Aspekt*' means phenomenological object, which appears to me this and that way. In another occasion, '*Aspekt*' is translated as 'the way of looking'. But the German original dramatically captures what Wittgenstein purports to say by the word. He says,

> It might consist in such facts as these: my glance doesn't move restlessly (inquiringly) around the object. I don't keep changing *the way I look at it*, but immediately fix on one and hold it stay.[6]

And the above is the translation of the following:

> Es könnte darin bestehen, daß mein Blick auf dem Gegenstand nicht unruhig (suchend) umherschweift, daß ich *den Aspekt des Gesehenen* nicht wechsle sondern sogleich einen *Aspekt* ergreife und festhalte.

In the above quotes, the English translation says "the way I look at it [object]." But we should consider its German original, "*daß ich den Aspekt des Gesehenen nicht wechsle* . . .," which clearly shows that "*den Aspekt des Gesehenen*" is "the aspect of what is seen," i.e., the (phenomenological) aspects of the (physical) object. So the translation might have been better off with "the way the object appears to me" than "the way I look at it [object]." A couple of more cases where Wittgenstein uses '*Aspekt*' as meaning 'appearance' appear in the *Philosophical Grammar*.[7] But a crucial, though indirect, evidence which unmistakably proves the phenomenological aspects of his philosophy and his use of '*Aspekt*' appears in Waismann's "Theses," where an attempt to interpret some of the main principles of the *Tractatus* is undertaken.

According to the "Theses," when we see an object, "what we observe are always only particular cross-sections" and these particular cross-sections are aspects.[8] So what appears to us in given visual experience is not the physical object itself, but the particular aspects of it. And the reason we can say that it is an object is because we add a hypothesis to what is given in our experience (i.e., aspects or a number of cross-sections). Waismann's "Theses" continues:

Particular aspects are what is changing and unstable; it is the form of the connection of aspects that is unalterable and remains. That unalterable connection is signified by *one* word.[9]

Thus, the particular aspects given in our visual experience, indeed, are phenomenological objects which Wittgenstein once believed can be picked out by 'this' and 'that'. However, 'this' and 'that', or the particular aspects cannot be signified by one word without the aid which obtains the greater simplicity by relying on a variety of hypotheses. This point also makes perfect sense when we remember Wittgenstein's phenomenological language is pure without any hypotheses, in contradistinction to a physicalistic language with hypothesis. Since Wittgenstein does not believe in pure phenomenological language any longer, the above quote saying we can signify the object by a word by means of connecting a number of aspects with the aid of hypothesis should not sound strange.

One more thing we should take notice in Wittgenstein's use of 'aspect' is that he does not confine his use to the visual experience. The sounds of music I hear can also be 'aspects', according to Wittgenstein. In the *Last Writings*, for instance, he says "In aesthetics isn't it essential that a picture or a piece of music, etc. can change its aspect for me?"[10] So Wittgenstein himself proves that his use of the word 'aspect' is unique. Although he does not develop his discussion on the aspect of sound fully, there is no doubt that he could have done so as well, as he actually mentions the parallel case to the visual experience in one occasion:

I could say of one of Picasso's pictures that I don't *see* it as human. Or of many other picture that for a long time I wasn't able to see what it was representing, but now I do. Isn't this similar to: for a long time I couldn't hear this as of a piece, but now I hear it that way. Before, it sounded like so many little bits, which were always stopping short—now I hear it as an organic whole. (Bruckner).[11]

There can be many different aspects of a visual object. And similarly, Wittgenstein argues that there can be different aspects of the same sound. A musical piece I could not recognize as an organic piece by a composer at one time can appear to me as a piece by Bruckner at another. When I could not hear it as Bruckner's piece, I might have heard the sounds unpleasant. But now I do not as I hear it as Bruckner's masterpiece. Wittgenstein could have said as well at this point that aspect-hearing is a parallel case to aspect-seeing. And regardless of the fact that there are not many places where he discusses on the change of aspects in auditory experience, it is undoubtedly true that he has this idea of aspect-hearing in connection with his phenomenology from the middle period on. The passage I quoted from the *Last Writings* above is not new to Wittgenstein, but one of the on-going problems he had with regard to his phenomenological ideas. A decisive case supporting this point appears in the material whose title is "Phenomenology is Grammar" in the Phenomenology chapter of TS 213. After briefly mentioning "Isn't the theory of harmony at least in part phenomenology and therefore grammar! The theory of harmony is not a matter of taste," Wittgenstein goes on to say,

> Understanding of a Gregorian mode doesn't mean getting used to the sequence of notes in the same sense in which I can get used to a smell and after a while cease to find it unpleasant. No, it means hearing something new, which I haven't heard before, much in the same way—in fact it's a complete analogy—as it would be if I were suddenly able to see 10 strokes ||||||||||, which I had hitherto only been able to see as twice five strokes, as a characteristic whole. Or suddenly seeing the picture of a cube as 3-dimensional when I had previously only been able to see it as a flat pattern.[12]

Wittgenstein here not only relates the case of auditory experience to that of visual experience, he says that recognizing a musical piece as a Gregorian mode means 'hearing something new'. It is indeed, what Wittgenstein's later self would love to call 'new aspect'. But more than anything, this aspect-hearing is, by Wittgenstein's own word, phenomenology.

As has been argued thus far, when Wittgenstein says the aspects 'this' and 'that' cannot be names, he undoubtedly criticizes his earlier self who believed 'this' and 'that' refer to phenomenological objects. But he no longer holds his earlier belief. That Wittgenstein no longer wants to consider 'this' and 'that' names appears in one of the earlier entries of *Philosophical Investigations*, where he critically examines his earlier views, and proves his consistency with what he says in the *Brown Book*. This time, Wittgenstein proposes that we should not call 'this' and 'that' names at all:

> But what, for example, is the word 'this' the name of in language-game (8)[13] or the word 'that' in the ostensive definition "that is called . . ."?—If you do not want to produce confusion you will do best not to call these words name at all.—Yet, strange to say, the word 'this' has been called the only *genuine* name . . .[14]

Again, the problem here is when we say 'this', the word can mean many different things or many different phenomenological objects, meaning of which cannot be successfully shown to others. The word 'this' relies on phenomenological (perspectival) identification which Wittgenstein had rejected together with the primacy of phenomenological languages. Not only that, Wittgenstein thinks of the possibility that one can even be confused by the identity of a phenomenological object (i.e. of the appearance of an object) in different times. So he asks,

> When I said to myself "What at one time appears to me like this, at another . . .", did I recognize the two aspects, this and that, as the same which I got on previous occasions? Or were they new to me and I tried to remember them for the future occasions? Or was all that I meant to say "I can change the aspect of this figure"?[15]

He indeed means phenomenological object by the word 'aspect' here, and even anticipates what in later writings comes to be the problems of seeing-as and aspect-change. Here he makes it clear that aspect is what appears to me in my experience. It is phenomenological object. We can have two or more phenomenological objects of one actual physical figure. One physical object can be given to us differently according to the viewing conditions of different times and places. Capturing phenomenological objects in our language is the general problem Wittgenstein had in his philosophy. What he thought was the solution in his early period turned out flawed as he realized the solipsistic pointing gesture is hung in the air. Language, for him, is no longer phenomenological. Words like 'this' and 'that' can only be meaningful when backed up by public means of identification. And language-games constitute the ground of all meaning. 'This' and 'that' will do all right when their use is mediated by language-games.

Now, an interesting, or even strange character of 'aspect' is that an object can also appear differently when the object is seen without the change of observing conditions. Wittgenstein was aware of this special case of the general problem from the Tractarian period. His example of the Necker Cube in the Tractatus, and that of a duck-rabbit picture in the later writings are the cases in point. Commenting on the Necker Cube, Wittgenstein says,

> To perceive a complex means to perceive that its constituents are related to one another in such and such a way.
>
> This is no doubt also explains why there are two possible ways of seeing the figure [the Necker Cube] as a cube; and all similar phenomena. For we really see two different facts.[16]

He might as well have said we really see two different aspects in the above passage. The actual drawing of the Necker Cube is one figure. But we can experience two differently oriented cubes from one and the same figure. These two different experiences come from seeing two different facts, as Wittgenstein puts it. Since there is only one physical figure, these two different facts are phenomenological aspects, i.e., two different appearances presented by an object to the eye. That Wittgenstein's 'facts' here are phenomenological supports the idea that the whole point of the *Tractatus* is about immediate experience and the immediate description of it. When he says proposition is a picture of fact, it means phenomenological language describing phenomenological fact. And the Necker Cube example is also consistent with the examples of the above-quoted use of '*Aspekt*' in the *Philosophical Grammar*, namely the way of appearance.

When Wittgenstein believes that there is phenomenological language describing phenomenological objects, phenomena like the Necker Cube are not a problem for him. But if phenomenological language is not available, it can cause a problem because in a physicalistic language there seems to be no way of expressing two different aspects of the Necker Cube. In other words, a drawing of the Necker Cube is a physical figure, and our language too is a part of the physicalistic system. Then how can we possibly express two different aspects, which are phenomenological, bearing only one physical construction? Instead of picking-up his old example of the Necker Cube, Wittgenstein develops a whole new case such as a duck-rabbit picture with a fashionable terms of 'dawning of aspect' or 'aspect-change'. This is also where the discussion of Gestalt psychology makes its appearance. But we have to be careful not to be misled by his discussion of Köhler. We should remember, from the entire texture of Wittgenstein's philosophical development, the late treatment of aspect is only a special case of the more general problem. As I mentioned, 'aspect' for Wittgenstein is phenomenological object. Now, his new problem is how we would express 'aspect' without phenomenological language.

2. LANGUAGE-GAMES OF ASPECT

Describing different aspects becomes a problem for Wittgenstein only after the rejection of phenomenological languages, that is, after the shift of his language paradigms. Wittgenstein now has to show how we can describe immediate experience in physicalistic language, for the rejection of phenomenological language does not mean the rejection of immediate experience at all. It might be said that the later Wittgenstein's move is to attack any view saying there is such a pure immediate experience in and of itself, for the later Wittgenstein appears to believe that we cannot make any direct description of direct experience without being mediated by language-games. Hence it might be said that his primary concern is no longer immediate experience or the dimension of it. Nevertheless, as I argued in the preceding section, it is very important to notice that his late discussion of aspect-seeing does not come out of vacuum, but is a left-over from much earlier stage of his philosophical development, in which we can explicitly set up its connection with Wittgenstein's own use of the word 'phenomenology'. So it may be true to say that there is no phenomenology found in the later Wittgenstein. But that does not exclude the possibility that he was indeed baffled with the problem of immediate experience, i.e., the problem that occupied him around 1930. He once believed immediate experience can be described directly. Now in the late 1940s, he does not do so. But that does not mean what drives his philosophical problems has essentially changed from the middle period. We can very well say that one of his concerns is how to cope with the problem of the description of immediate experience. He needs to say something about how we are able to talk about what we are immediately aware of in our one and only language, which is an everyday physicalistic one.

In other words, there can well exist phenomenological objects in Wittgenstein's post-Tractarian philosophy, although speaking of such objects directly is impossible. He simply cannot spell out what the examples of phenomenological objects would be like, as he could in the *Tractatus*. But his silence should not be mistaken for the denial of such existence. A duck-rabbit picture is one example presented with the same spirit of the Necker Cube. In fact, we can actually find many cases from the middle period on, where Wittgenstein deals, if not struggles, with the problem of the way we describe immediate experience. The so-called private language argument is one well-known case, and the treatment of color concepts is another.

In the pain case, children's learning of pain-language is the process of replacing crying or exclamations. When a child hurts himself and cries, crying is the most direct expression of his pain. It is the primitive expression of pain. But an adult teaches a child how he can express his pain in verbal language, such as "I am in pain." Now crying is replaced by the word 'pain'. But it neither means that there is no such thing as pain any longer, nor the word 'pain' means crying (*PI*, sec. 244). It only means the replacement of one pain-behavior (natural, primitive) by another (verbal). And this new pain-behavior cannot be private or phenomenological, though pain itself may well be; for pain-language is learned by the mediation of the public markers such as grimace and other facial expressions. Thus, even pain can be expressed in physicalistic language. It is mediated by the public language-games of pain with the aid of physiognomic expressions.

We can discuss the color case in a similar fashion. The way we use color-words is equally spontaneous as the way we use pain-words. We cannot make as it were drive a wedge between the color impressions we have and the color-words we use. But that does not mean the color-words operate in the way in which we put a label on our perception. The experience of color can be expressed only by depending on the physical framework of language.

Now, Wittgenstein has the case of aspect at hand, whose expression is also spontaneous as the pain or the color case. To say the conclusion first, Wittgenstein's point is that expressions of aspect "Now I see it as this" or "Now I see it as that" are possible only by means of relating those aspects to the relevant public language-games. Of course, Wittgenstein would probably protest against this view on the basis that the way we get pain-expressions is not entirely the same as the way we get aspect-expressions. We can say that a pre-linguistic child is in pain regardless of the fact that he has not learnt any language yet, and thus is not able to express his pain in verbal behavior. His facial and other primitive expressions will convince us of his being in pain. On the contrary, there are no other ways to express the experience of aspect than our language-games of aspect. We may say 'this' and 'that' are primitive forms of expression in the aspect case, but these expressions differ from the primitive expressions of pain, as they do not really show what experience they mean. Therefore, the pain case and the aspect case are not the same.[17] But we can see the related ideas that is common in both cases, namely we are trained to behave in certain ways, and thus master certain techniques of expressing experience. So we can safely say that the nature of problem and solution in the case of aspect is almost the same as the earlier case of 'pain' or 'color'. Wittgenstein's own writings support this point. In the *Last Writings*, Wittgenstein briefly discusses aspect-changes in the earlier entries (secs. 165-180). But instead of a prolonged discussion which only comes out in much later entries, his overview on aspect-change here leads his discussion to the expression of pain (secs. 183-184, 203-205, and 236-243) and color (secs. 207-208, 213-222). In between, there are a couple of physiognomic episodes (secs. 201, 202).

In all of the occasions, Wittgenstein's discussion leads itself to the direction of games played publicly. He declares that one's expression of an aspect is not a perceptual report (*LW*, i, secs. 176, 486). We can easily see the point of this, as pain-words or color-words are not perceptual reports. And we can see that the description of an aspect can only be made when one relates one's perceptual experience to the concepts that involve physical attributes. Knowing whether another person is in pain, too, is based on the game we play with a tool (*LW*, i, sec. 243, 247). A thought-experiment of the people who use color-words such as 'leaf-color' instead of red and green, too, implies the language-games played with physical attributes (*LW*, i, sec. 220). And the physiognomic episodes inserted in between show us that the very public nature of facial expressions, and the mastery of them enables us to communicate what goes on in our immediate experience (*LW*, i, sec. 201). Therefore, it is obvious that Wittgenstein puts the problem of describing aspect on the same level as the problem of pain-words or that of color-words.

Perhaps the simplest example that shows what Wittgenstein's problem is in aspect-seeing is the case when we have a vague image in our visual field. When something is given in one's visual experience, what would be the most immediate description of that experience? Indeed, it is a question of the same sort as "What would be the most immediate description of pain when a child hurts himself?" A child would cry as discussed earlier. Now what one would say when one sees a vague figure in one's visual field? One would be really tempted to say simply "This." But since the middle period on, 'this' is no longer an official name of an aspect. So in various examples, Wittgenstein tries to show the possible ways in which one can relate one's visual experience to some sort of physical attributes. When one sees a vague figure 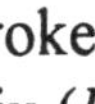 , one can describe it as the alphabet T with an additional stroke (*RPP*, i, sec. 533), as a combination of a T with a little appendix (*RPP*, i, sec. 539), or just as an F (*RPP*, i, sec. 510). In any case, the point is that in order to express what we experience we somehow relate our experience of aspects to the publicly knowable figures like T or F. So basically, Wittgenstein's problem here is how we can successfully express aspects in our physicalistic language. But as his very late writings show, there is more special case of the problem, and it is the case like a duck-rabbit picture.

What makes a duck-rabbit picture an interesting case is that we can see it as a duck at one time and as a rabbit at another without optical or any other physical change in the picture itself. There is one and the only picture, but there are changes in visual experiences. Since the physical figure of duck-rabbit remains the same, the aspects cannot be ascribed to any physical properties of the figure (*LW*, i, sec. 516; *PI*, p. 212). To explain this change in visual experience, Wittgenstein introduces the term 'dawning of aspect' (*LW*, i, secs. 429, 433; *PI*, p. 193). So it looks as if Wittgenstein was interested in this strange psychological phenomenon itself; that is, its cause, mechanism, etc.. But he makes it clear that he is not interested in the cause of the sudden change of aspect (*LW*, i, sec. 434; *PI*, p. 193). Rather, what he is concerned with is phenomenological differences we have when we see a duck-rabbit picture. Thus he states that what appears to the perceiver when one sees a drawing is not physical construction of the drawing. One may think what one experiences cannot directly be shown to others. It simply indeed reminds us of the pain case when Wittgenstein's interlocutor says, "After all, my visual impression isn't the *drawing*! It is *this*, which I can't show to anyone." (*LW*, i, sec. 440; *PI*, p. 196) Clearly Wittgenstein opposes the interlocutor's assertion on the same ground as the pain case in which someone claiming that he knows what pain is only from his own case, but does not know whether other people also have pain (*PI*, sec. 293). One can successfully show one's pain to others, as one can talk about sensations everyday and give them names (*PI*, sec. 244). If one says "It is *this*, which I can't show to anyone" when one has visual experience, the depiction of aspect by 'this' is solipsistic expression, as criticized by the middle Wittgenstein. 'This' is not the name of visual object. That kind of expression should be mediated by language-games.

But it is also clear that Wittgenstein thinks 'this' and 'that' are the expressions of aspects. They are indeed the most immediate descriptions of aspects in his early philosophy. If 'this' and 'that' are not used solipsistically, and mediated by public language-games, their expressions are fine. Wittgenstein says,

> There are many ways of experiencing aspects. What they have in common is the expression: "Now I see it *as that*"; or "Now I see it *this* way"; or "Now it's *this* – now *that*"; or "Now I hear it as . . .; a while ago I heard it as . . ." But the explanation of these "that's" and "this way's" is radically different in the different cases.[18]

This passage is also similar to the passage from the *Philosophical Investigations* I quoted in the preceding section. There Wittgenstein points out that 'this' is not a name at all, for it can mean many different things. Without providing the way we specify the categories to which an object belongs, 'this' will be obscure in meaning to others. The similar, if not the same, thing happens in the above quote. 'This' and 'that' can be used to designate different aspects (of seeing or hearing). But the direct description of aspect like "I see it like *this*" pointing to the same thing for 'it' and 'this' cannot convey what one sees in one's visual experience to others.[19] It is solipsistic indeed and the pointing gesture does not help as the pointing hand also belongs to the visual experience. The expression lacks public means of identification.

Wittgenstein's solution to this problem is to bring in the idea of language-games. The very reason that we can say "Now I see it as duck" or "Now I see it as rabbit" is because we are so trained that we are able to relate our aspects to many different actual things such as duck or rabbit. In other words, we can play the language-game of aspects. So we learned the words and expression for aspects. And we learn them from samples or other physical correlates. In a duck-rabbit picture, we learn how to distinguish two different aspects only in the language-games of aspect, initially with the help of the pictures of duck and rabbit or the actual duck and rabbit. For otherwise there would be no way to express two different phenomenological aspects in our language. Simply having the sense-impressions of the two different aspects from a duck-rabbit picture would not enable us to express the two different aspects in our language. What makes us speak "I see it as rabbit" or "I see it as duck" is the language-game of aspects which we mastered by relating those impressions of aspects to the pictures of rabbit and duck. And sometimes we may really need to point to real rabbits or real ducks (*LW*, I, secs. 466, 532).

We learn the language-games of aspect. The same drawing can be seen as rabbit or as duck. But it does not matter, and we can express different aspects of the drawing, for we are so trained in language-learning that we can distinguish and express the different aspects by relating them to the language-games of aspect. We learn the language-games of aspect, and we learn how to behave in case we have such and such experience. In other words, we learn a skill or technique of verbal behavior to express the experience of aspects. Wittgenstein says,

> It is only if someone can do, has learnt, is master of, such-and-such, that it makes sense to say he has had this experience.[20]

The above quote is from where Wittgenstein deals with aspect-seeing. But the idea can well applied to the pain case where language-learning plays an important role. Thus, in the pain case we can say "He is in pain," and 'his pain' here must be of the same kind as our pain. The reason we can say his pain is the same as ours is that he—as well as we—has learnt, and is able to behave in certain ways when in pain. And this is not essentially different from the case where we make a statement regarding different aspects, as the above and the following passages vividly show:

> "Now he's seeing it like *this*", "now like *that*" would only be said of someone *capable* of making certain applications of the figure quite freely.
>
> The substratum of this experience is the mastery of a technique.[21]

Now we can see the characteristic of aspect-game is not essentially different from that of pain-games. And this characteristic of aspect-games is not much different from the language-games of pain where imagining expression-charts can play a role. Our using pain-words means the replacement of one pain-behavior (primitive) by another (verbal). The same point indeed is made in the aspect case.

> I want to say that the natural, primitive expression of the experience of an aspect would be such an exclamation.[22]

Both things, both the report and the exclamation, can be called expressions of perception and of visual experience. But the exclamation is so in a different sense from the report; it is related to the experience as a cry is to pain.[23]

The very expression which is elsewhere a report of what is seen, is here a cry of recognition.[24]

So the verbal expression of aspect replaces the primitive expressions of aspect, as the verbal expression of pain replaces the primitive expression of pain. In the pain case, a child, before he learns pain-language, would cry. And the expression "I'm in pain" is the verbal expression that replaces crying. And this verbal one is the language-game we learn. In a similar fashion, the aspects would be described as 'this' and 'that'. In particular, if a child really has not seen any pictures of rabbit or duck, there would really be no other way to express what he sees than 'this' and 'that'. And still, 'this' and 'that' will not say anything, because they do not distinguish and show any difference in aspects. It is only after one is able to relate what one sees to the physical attributes such as picture-duck or picture-rabbit when one can use the expressions of aspects in such a way that listeners can make sense out of the expressions. And this is what the series of quotes above amount to illustrating.

Therefore, we can say that Wittgenstein's extensive examination of aspect-seeing in his late period is not primarily intended to discuss psychological phenomena we have and their concepts. Also, it is not the case that the underlying motivation of Wittgenstein's discussion of aspect-seeing is to react against Gestalt psychology. It is true that he often discusses the examples presented, and the points made, by Köhler. But it does not mean that Wittgenstein's main concern is Gestalt psychology and its theme. He tackles Köhler's examples because they give challenging problems regarding the expressions of immediate experience. No matter what kinds of phenomena there are, Wittgenstein had to explain that we should be able to express them in our language, that is language-games in and by which phenomena are related with physical attributes.

It is evident, from what I have argued, that the later Wittgenstein's discussion of aspect-seeing is the problem existed from the middle period. The problem of aspect has been persistent one from when

Wittgenstein explicitly relates it to phenomenology. And his later discussion of aspect only proves that the phenomenological theme pervade his entire work in philosophy.

CHAPTER VI

IS PHENOMENOLOGY POSSIBLE FOR WITTGENSTEIN?

1. OTHER PHENOMENOLOGICAL INTERPRETATIONS

There have been intermittent attempts to view Wittgenstein's philosophy as phenomenology since the 1950s. In the 50s and 60s, with a virtual inaccessibility to the vast amount of the *Nachlass* materials, most attempts were limited to simple comparison of the themes and method of Wittgenstein's philosophy with those of Husserlian phenomenology. Although most philosophers who belonged to this trend failed to identify Wittgenstein's philosophy with phenomenology in any explicit manner, their contribution to the phenomenological interpretation of Wittgenstein's philosophy was important in the sense that they were the first to group Wittgenstein and Husserl together; for it means that they, at least, sensed the elements of phenomenology in Wittgenstein's philosophy.

Perhaps the first paper that took this stand was written by van Peursen. In his 1959 paper entitled "Edmund Husserl and Ludwig Wittgenstein,"[1] van Peursen presents a careful survey on similarities and differences between Husserl and Wittgenstein, not endorsing any view that claims Wittgenstein's philosophy is phenomenology. One significant contribution made by van Peursen was his pioneering approach of viewing the methods of philosophy employed by the two prominent philosophers as sharing similarities in such a way that they undermine any traditional sense of metaphysics. Not only that, van Peursen compares the common themes between the two, showing the apparent differences between them.

The fact that van Peursen came close to grouping Wittgenstein and Husserl together is significant in the sense that he at least sensed some similarities between these two philosophers, who were believed to represent the two different traditions of the philosophical scenes.

But his survey falls short of identifying Wittgenstein's philosophy as phenomenology in any explicit manner. None of Wittgenstein's texts containing clues to phenomenological interpretation were published in the 1950s,[2] and van Peursen relied for his view entirely on the *Tractatus* and the *Philosophical Investigations*, which resulted in an interpretation missing the crucial link that would explain Wittgenstein's real philosophical concerns, the difficulties he faced, and the solution he finally came up with. Wittgenstein's main philosophical concern from the earliest stage of his career was immediate experience and how to describe it. But this problem can be best viewed through his early middle-period writings such as the *Philosophical Remarks*, and the related works like *Ludwig Wittgenstein and the Vienna Circle*. These works are extremely important in the phenomenological reading of Wittgenstein's philosophy, not only because the words 'phenomenology' and 'phenomenological' appear in those works, but because the problem of immediate experience and the role and function of language in describing immediate experience are present in them. Thus, these middle-period works enable us to understand the right context in which Wittgenstein's line of thoughts stands. Also, they are important because of Wittgenstein's silence regarding the phenomenological theme in the *Tractatus*. But with the middle-period evidence, we can at least reconstruct Wittgenstein's earlier problems. This part is absent in van Peursen's survey, which prevents him from an appropriate and comprehensive understanding of Wittgenstein's phenomenology.

Another case that can be classified as this early approach is represented by Ricoeur, who also attempts to compare the position and function of Wittgenstein's philosophy to Husserl's phenomenology. Ricoeur in his 1967 paper "Husserl and Wittgenstein on Language"[3] claims that the idea of representation of possibility in Wittgenstein's *Tractatus* comes close to phenomenology. To be sure, Wittgenstein's idea of representation of state-of-affair is not that of what is, but of what can be, for it deals with possibility. In this sense, Ricoeur's view is that phenomenology occurs in Wittgenstein's philosophy. This is going one step further than van Peursen toward the phenomenological interpretation of Wittgenstein's philosophy. However, Ricoeur immediately steps backward from this and says that "this phenomenology tends to abort."[4] To him, Wittgenstein's phenomenology cannot be made explicit, because his picture theory lacks in reflectivity. While a picture can depict any reality whose form it has, it cannot depict its own form. Ricoeur believes that this lack of reflectivity excludes any possibility of phenomenological theory in Wittgenstein, for what makes one fact possible and another impossible cannot be reflected on a discourse. Thus, Ricoeur's view is that Wittgenstein's position in the *Tractatus* remains as a theory of logical picture which is concerned with logical forms.

Ricoeur's case is also important in the sense that he discovered the elements of phenomenology in Wittgenstein's philosophy. But he, too, fails to identify Wittgenstein's philosophy as phenomenology in an explicit manner. Although the *Philosophical Remarks*, where Wittgenstein frequently uses the term 'phenomenology' or 'phenomenological' was published in German in 1964, Ricoeur doesn't seem to have read the book when he wrote the paper. Obviously, with this lack of understanding in Wittgenstein's middle-period works, it is not too difficult to understand why Ricoeur fails to grasp Wittgenstein's consistent problems with phenomenology. For the alleged lack of reflectivity cannot be questioned meaningfully if we understand Wittgenstein's philosophical context correctly. Wittgenstein's problem is immediate experience and how to describe it. The language that would describe what is given in immediate experience will itself be the ultimate medium that links language and the world. And he tries to find the whole structure of reality in our language, which itself plays the role of universal medium. So he, at first, tries to construct an ideal language that would mirror the world directly. The middle period, if it existed, is the stage where Wittgenstein faces the problem arising from this ideal picture and begins to change his view on language paradigms. But this does not change his firm belief in language as the ultimate medium that will function as a link to the world, as his much celebrated idea of language-games serves this purpose through his late philosophy. In this way, Wittgenstein's concern is directed toward language, in which he believes all the phenomenological problems can be treated. Since we speak through our language, Wittgenstein believes language itself cannot be the object of discourse; semantics is ineffable. Thus, even though he speaks of the possibility of sense, the ineffable nature of semantics necessarily results in a different kind of phenomenology from Husserlian phenomenology.[5] The problem of reflectivity cannot be explicitly present in this sense. However, that Wittgenstein also deals with the phenomenological problems cannot be denied just because he does not show the elements of reflectivity explicitly. Pointing out the elements which are parallel to each other is important. But what should not be asserted is that only Husserlian phenomenology is phenomenology. With no trace of interaction between Husserl and Wittgenstein, it would be very unlikely that Wittgenstein did Husserlian phenomenology all by himself. Rather, what we should ask first is, regardless of the differences in philosophical methods and vocabulary, whether Wittgenstein really had the same problems as Husserlian phenomenology, and if he did, how he approached to these problems for possible solutions.

The second group of philosophers who attempted to interpret Wittgenstein's philosophy from the phenomenological standpoint emerged from the late 60s through the 70s and the early 80s. Most of these philosophers were much helped by von Wright's pioneering studies of the *Nachlass* materials and the subsequent publications from the *Nachlass*.[6] When the *Philosophical Remarks*, where Wittgenstein's words 'phenomenology' and 'phenomenological' make their appearances, was published, scholars paid special attention to Wittgenstein's use of the words and tried to identify Wittgenstein's philosophy from the phenomenological viewpoint. Hence, it would be reasonable to say that the full-fledged analysis of Wittgenstein's philosophy from the phenomenological perspective only began after—at least—some of the *Nachlass* materials were published. Having read that Wittgenstein in fact used the term 'phenomenology' or 'phenomenological language', some philosophers began to turn their interests to whether Wittgenstein was really a phenomenologist. Further, some have made attempts to relate Wittgenstein's phenomenology to Husserlian phenomenology as well as existentialism. Herbert Spiegelberg and Nicholas Gier are representatives of those who belong to this group.

Spiegelberg is indeed the first to systematically view Wittgenstein's philosophy from the wide perspective of the phenomenological movement. Spiegelberg's account is particularly different from those mentioned above in the sense that he carefully studied Wittgenstein's *Nachlass* materials and tried to figure out what Wittgenstein really wanted to say with his term 'phenomenology' or 'phenomenological'. Also, he tried to diagnose whether Wittgenstein's use of the terms and underlying themes can be considered as belonging to the phenomenological movement initiated by Husserl. In his 1968 paper "The Puzzles of Wittgenstein's *Phänomenologie* (1929-?)" and its 1979 supplement, Spiegelberg argues that Wittgenstein's phenomenology was generated in 1929 and developed through the writings of the Big Typescript in 1933, but that it was completely replaced by the grammar of ordinary language beginning with the *Blue Book*.[7] For Spiegelberg, no phenomenological elements in Wittgenstein's philosophy seem to exist before 1929, and the reason for the disappearance of Wittgenstein's use of the term 'phenomenology' after 1933 remains a puzzle.

Regardless of the study of *Nachlass*, Spiegelberg fails to do justice to Wittgenstein's early philosophy, where the most important characteristic of phenomenology is present, because he asserts that Wittgenstein's phenomenology, if any, began in 1929. Since Wittgenstein stopped using the words 'phenomenology' and 'phenomenological language' after the mid-30s, Spiegelberg's attempt to investigate Wittgenstein's phenomenology on the basis of the appearance of the relevant terms faces a puzzling question of when and why Wittgenstein gave up phenomenology, which cannot be answered successfully. In fact, Spiegelberg conducts a very careful investigation of Wittgenstein's use of the word 'phenomenology' or 'phenomenological', careful enough to count how many times the word appears in the *Philosophical Remarks*. Furthermore, as new materials from the *Nachlass* were studied and published, Spiegelberg went on to publish "Supplement" to his original 1968 paper. Nevertheless, the important thing that Spiegelberg missed is an attention to the context and relation of Wittgenstein's middle-period writings with regard to the early works, including *Notebooks* and the *Tractatus*. Even in his supplement to the original paper published in 1979, while he was crediting Wittgenstein's abiding interest in phenomenology upon the discovery of his use of the words 'phenomenology' and 'phenomenological problems' in *Remarks on Colour*, Spiegelberg failed to suspect that Wittgenstein might have dealt with the phenomenological problems throughout his entire philosophical career and did not consider the possibility that Wittgenstein was a phenomenologist from the beginning. It is rather surprising that Spiegelberg never mentioned the possible influences of phenomenological physics on Wittgenstein as it has been well known that Wittgenstein had enough knowledge of Boltzmann, Hertz, Mach, and other theoretical physicists, because Spiegelberg himself was aware of the fact that Husserl and Mach had some form of philosophical interchanges and the former had an interest in the method of phenomenology employed by the latter.[8]

Regardless of all the efforts to figure out what Wittgenstein really wanted to say with his use of the words 'phenomenology', 'phenomenological language', and 'phenomenological problems', Spiegelberg, too, falls short of identifying Wittgenstein as a phenomenologist who was almost obsessed throughout his entire philosophical career with the problem of immediate experience and how to describe it. In a critical manner, he aptly points out that Wittgenstein never gives any definition of his word 'phenomenology' or 'phenomenological language', or that his use of the word 'grammar' is unclear and ambiguous throughout his writings. Certainly, Wittgenstein never formulates his idea of phenomenology in any of his surviving writings. Perhaps, the clearest example in which he asserts something about phenomenology would be one of the chapters in the Big Typescript which is entitled "Phenomenology is Grammar." It is true that this is not clear enough for most philosophers to understand what Wittgenstein really means by phenomenology, for he never gives an explicit explanation of what he means by the word 'grammar' either in the chapter in question or in any other writings. But this lack of clarification in Wittgenstein's vocabulary cannot be used as a point to disqualify any phenomenological readings of Wittgenstein's philosophy, because clarifying philosophical terms is certainly not one of his philosophical habits. For example, if we consider the fact that Wittgenstein never gives us a clear definition of language-games, it is no wonder that he did not feel like giving any clear formulation of his much concerned ideas of phenomenology. Even the word 'picture theory' was an invention of later philosophers. And if we doubt Wittgenstein's use of the word 'phenomenology' because of the lack of clarity or definition, there is no reason for us not to doubt his use of other major ideas such as 'language-games' which are there in his writings but hardly defined or clarified.

Another example of the phenomenological interpretation that ignores the phenomenology of the *Tractatus* is found in Nicholas Gier's *Wittgenstein and Phenomenology*. Gier goes much further than Spiegelberg by asserting not only that Wittgenstein was a phenomenologist from 1929 to the end of his life, but his later philosophy could be meaningfully compared with other philosophers who stand in the phenomenological tradition such as Heidegger and Merleau-Ponty. One of Gier's reasons for not viewing the *Tractatus* as a phenomenological work is that he believes Tractarian objects cannot be identified with the phenomena of immediate experience.[9] Also, he argues that "Wittgenstein's commitment to external relations, atomism, and representationalism definitely separates the *Tractatus* from the phenomenological movement."[10] Thus, to Gier, Wittgenstein's early philosophy contains a rather metaphysical view on the simples which was given up around 1929 to develop his own phenomenological method.

Gier's assertion that Tractarian objects are the metaphysical simples cannot be supported from the context that Wittgenstein under the influence of Russell developed the modified theory of acquaintance in his *Tractatus*. Not only that, in understanding Tractarian objects, we should not ignore the Moorean point that differentiates the object of consciousness and the act of consciousness, which comes close to the idea of intentionality in Husserlian phenomenology. As was discussed in chapter 2, section 2 above, various evidence supporting the point that Tractarian objects are the objects of immediate experience can be found in many writings, not just in Wittgenstein's own words as the self-criticism of his earlier view, but also in the records of close philosopher friends. Perhaps, the most explicit statement regarding his own earlier view made by Wittgenstein himself is section 46 of the *Philosophical Investigations*. This section was partly quoted earlier, but is worth quoting here again to emphasize the close tie between Wittgenstein and Russell and what comes out of it:

> What lies behind the idea that names really signify simples?—
>
> Socrates says in the Theaetetus: "If I make no mistake, I have heard some people say this: there is no definition of the primary elements—so to speak—out of which we and everything else are composed; for everything that exists in its own right can only be *named*, no other determination is possible, neither that it *is* nor that it *is not* . . . . . But what exists in its own right has to be . . . . . named without any other determination. In consequence it is impossible to give an account of any primary element; for it, nothing is possible but the bare name; its name is all it has. But just as what consists of these primary elements is itself complex, so the names of the elements become descriptive language by being compounded together. For the essence of speech is the composition of names."
>
> Both Russell's 'individuals' and my 'objects' (*Tractatus Logico-Philosophicus*) were such primary elements.[11]

What this long passage show is that, in fact, Tractarian objects are identified with Russell's simples, which he termed objects of acquaintance and the typical example of which is sense-data. Russell's sense-data not only serve as logical atoms in our conceptual structure, but they also impinge upon our minds. Sense-data are not mental. Thus, identifying Tractarian objects with the objects of immediate experience, we can get away from the danger of phenomenalism. Since the objects of immediate experience are equipped with logical forms, the possibility of figuring out the structure of reality wholly lies in our immediate experience. Unlike Gier's objection, this view supports the idea that the *Tractatus* in fact contains phenomenological themes. And this earlier position lasts through the middle period even after Wittgenstein himself declared the impossibility of phenomenological language. Wittgenstein's remark supporting this is once again worth quoting:

The world we live in is the world of sense-data; but the world we talk about is the world of physical objects.[12]

There would probably be no better example than this to show what Wittgenstein's main concern was in the early and the transition periods. In the early period, he was concerned with the objects of immediate experience (sense-data) and a pure language that describes those objects (phenomenological language). In the middle period, what Wittgenstein came to realize is the impossibility of such an ideal language as independent notation. All we have is our everyday language, which is physical. But, while language is physical, Wittgenstein is still concerned with the world of sense-data, which is primary in the sense that we live in it. As a matter of course in this context, it would be very strange if early Wittgenstein's position was far different from this and made only a metaphysical point with regard to the simples. The phrase "the world of sense-data" in which we live is indeed a left-over from the early Wittgenstein, and should not be understood as a sudden breakthrough around 1929. The phenomenological problem is the abiding theme throughout the entire period of Wittgenstein's philosophy.

The position that Wittgenstein's phenomenology began in his middle period has led Gier to compare some fragments of ideas in Wittgenstein's later philosophy to Husserl's phenomenology. One example is Gier's contention that Wittgenstein in his later philosophy performs phenomenological reduction. With regard to Wittgenstein's firm view that the language of science cannot be used to describe true phenomena, Gier says "Wittgenstein has obviously accepted a type of phenomenological reduction similar to Husserl's epoché."[13] According to Gier, the epoché is the linchpin of phenomenology's strictly descriptive method, a method which Wittgenstein enthusiastically endorses, quoting the following remarks as his supporting evidence:

We must do away with all *explanation*, and description alone must take its place. (*PI*, sec. 109)

Philosophy really *is* 'purely descriptive'. (*BLB*, p. 18)

But we must be careful in using these remarks to support the idea that Wittgenstein was doing phenomenology, or that he employed a method very similar to the method used in Husserlian phenomenology, because the quoted remarks were written well after Wittgenstein rejected phenomenological language, the function of which he believed was to describe immediate experience.[14]

To support that what Wittgenstein calls 'descriptive method' is parallel to the phenomenological reduction, Gier goes on to quote other passages from the *Philosophical Remarks*:

The worst philosophical errors always arise when we try to apply our ordinary—physical—language in the area of the immediately given . . . . All our forms of speech are taken from ordinary, physical language and cannot be used in epistemology or phenomenology without casting a distorting light on their objects.

Furthermore, it must be a theory in *pure* phenomenology in which mention is only made of what is actually perceptible and in which no hypothetical objects—waves, rods, cones, and all that—occur.[15]

Obviously, what Wittgenstein really talks about in the above-quoted passages is phenomenological language, which is the language of immediate experience, and of which the possibility and necessity are rejected in the end. This phenomenological language cannot be simply identified with the 'descriptive method', which Wittgenstein went on to talk about. Hence, we should rather interpret the above quoted passages to mean that Wittgenstein in his early middle period plays with the viability of phenomenological language which he has believed from the earlier period, in comparison with the notion of physicalistic language, which is the only language we have in his new philosophical development.

If, as Gier maintains, what Wittgenstein calls 'descriptive method' should be identical to something like phenomenological language, then that means Wittgenstein is reviving a once abandoned notion in a new terminological phrase. This is impossible because once Wittgenstein rejected the possibility of phenomenological language as our primitive language, he allows its operation only within the framework of physicalistic languages. As was discussed in chapter 2, section 2 above, there is no change of position for Wittgenstein with regard to the basic phenomenological ontology; i.e., the world we live in is the world of sense-data. But since the language we have is physicalistic, all talk of sense-data, too, is undertaken in a physicalistic language system. A particular case of sense-datum language can be possible if it depends on the framework of physicalistic language.

What Gier fails to see is the context of his quoted passages. It is not Wittgenstein's sudden realization or possible influence from Husserl in the middle period that gives him the motive to talk about descriptive method without hypotheses. Rather, it is deeply connected to Wittgenstein's long fought philosophical battle that phenomenological language should give up its place to physicalistic language. Gier seems to believe that this descriptive method, in contrast to the scientific method, consists of Wittgenstein's phenomenological reduction. But once again, this false idea results from the wrong assumption that Wittgenstein's phenomenology begins after 1929. Wittgenstein's phenomenological ideas prevail within the entire context of the *Tractatus*. And if his middle-period attitude against the scientific method and hypothetical elements is supporting evidence for his phenomenological reduction, then it should not be any different in the attitude of the early Wittgenstein where he deals with logical points in contrast to factual or scientific points.

In the *Tractatus*, objects have to be given in one's experience. Without this givenness, there is no logic. Without the givenness of objects in one's experience, no meaning in language is possible in Wittgenstein's early system. It is because Wittgenstein develops the whole point with regard to logic out of the disagreement with Russell's theory of judgment, where the separate class of logical forms to be acquainted is required. As a variation of, or even as a remedy to, Russell's theory, Wittgenstein comes up with the idea that our ability to have logic is given with objects, for objects to be acquainted are already equipped with logical forms so that the entire possibility of how objects are combined to form the world is already built into each object. Therefore, the entire logic and the meaning of our language, too, are already written into objects, which are the contents of immediate experience.

If there is anything that comes close to Husserlian phenomenology in this context, it is that Wittgenstein tries to figure out the entire conceptual structure from what is given in immediate experience. Undoubtedly, Wittgenstein has no interest in scientific methods or hypotheses in this regard. Even dealing with the problem of color-incompatibility, Wittgenstein does not mean physical incompatibility. The whole point throughout the *Tractatus* is a sort of logical possibility that the givenness of object of immediate experience brings about. Thus if, as Gier maintains, there were any epoché in Wittgenstein's philosophy, then it cannot be a feature that suddenly comes out in his middle-period writings. His middle-period concern is only meaningfully discussed when viewed as a continuation of the early period.

Another important point to be emphasized here is that we should not attempt to bring Wittgenstein's phenomenological ideas into a direct comparison with Husserlian phenomenology as if Wittgenstein was a Husserlian phenomenologist. With insufficient evidence about the philosophical relation between Wittgenstein and Husserl, which is unlikely to come up in the future, it is really not appropriate to try to make a point of one-to-one correspondence between the two different phenomenologists. The mere fact that Wittgenstein mentions descriptive method and the language of immediate experience would not make his approach akin to Husserl's phenomenological epoché. In a strict sense, there cannot be any phenomenological reduction for Wittgenstein, because what Husserl wants to achieve by phenomenological reduction can be achieved without much ado in Wittgenstein's case. What Husserl wants to achieve by the phenomenological reduction is to get at the immediately given. He wants to bracket out everything doubtful of consideration. The way we reach the real object and its meaning comes only after the phenomenological reduction is performed. So for Husserl, the phenomenological reduction is a way to get purity in search for the foundation of all knowledge. What we have after the reduction is pure consciousness that survives the reduction. But this pure consciousness is not the only thing we have. The reduction gives us the possibility of having reality within our consciousness, which was initially transcendent before the reduction. This is different from the case of Wittgenstein where the very possibility of having reality is simply given without the transcendental turn. And it is only because Wittgenstein's objects given to us are equipped with logical forms.

For Husserl, however, the possibility of having reality is not gained by simple acquaintance. In the course of the phenomenological reduction, according to Husserl, something previously transcendent becomes transcendental through the process of form-giving activity to the given hyletic data. It is ascribed to the intentional characteristic of our pure consciousness which makes this constituting process possible. The material given to us in the course of the phenomenological reduction is un-formed raw material, which alone cannot be called phenomenological objects. What is necessary in this case is to make this material meaning-entity. And the very activity of meaning-giving can only take place from the subject side of the human mind. Unlike Wittgenstein, there can be no purely empirical phenomenological objects for Husserl. All phenomenological objects for Husserl contain non-empirical elements in the sense that the forms are imposed upon by the noetic activity. Without doubt, this is the very point that keeps Husserl's phenomenology from pure realism. And also, this keeps Husserl from having much interest in phenomenological language that would immediately describe what is given in immediate experience, for he has to deal first with reduction and constitution in order to have proper phenomenological objects. But once we have phenomenological objects, the problem of language that would describe them would not come up as an important issue with the constraint of idealism in Husserl.

On the other hand, for Wittgenstein, this grounding does not require a special method. All we have to do is simply experience objects, and it is because what gives meaning to us comes together with objects. In contrast to Husserl's case, where the role of subject proves important, Wittgenstein's immediate experience is itself the basis in which every possible logic is given and from which the entire meaning-structure of the world can be built. No special operation to grasp reality is necessary, because the key to grasping reality, namely logical forms, is already given. In this regard, for Wittgenstein, no greater role of subject than to experience objects is necessary.

Therefore, Wittgenstein, having no phenomenological reduction, can have purely phenomenological objects captured in immediate experience. These objects are given with their logical forms, and therefore, are categorially well-defined. This unique feature enables Wittgenstein to discuss phenomenological language that describes these phenomenological objects in immediate experience. At least throughout the Tractarian period up to 1929, Wittgenstein believes there is this kind of language, and tries to construct suitable notations. And, without doubt, what makes this possible is Wittgenstein's belief that objects of immediate experience are given with logical forms. Since the transparent relation between language and object is set up, it becomes possible to speak of the immediate description of immediate experience.

Gier's attempt to find the phenomenological reduction in Wittgenstein is ill-founded in this regard. For Wittgenstein, the transcendental turn is not necessary. Nevertheless, if we still try to find any counterpart of the phenomenological reduction of Husserlian phenomenology in Wittgenstein, the closest possible example, perhaps, would be Wittgenstein's reduction to objects of acquaintance, the original idea of which was initiated by Russell. Because the possibility of how language and the world are structured can be found in simple objects, the importance of reduction to objects in Wittgenstein's phenomenology should match that of the phenomenological reduction in Husserl's phenomenology.

That Wittgenstein's Tractarian objects are given in experience together with logical forms entails another significant difference from Husserl's phenomenology, which Gier equally misses. It is the view on synthetic *a priori*. Gier contends that Wittgenstein changed his mind regarding his view on synthetic *a priori*:

Wittgenstein's surprising acceptance of a synthetic *a priori* is at odds with his forceful rejection of the concept in the *Tractatus*. Here mathematical propositions are tautologies, i.e., analytic truth that show nothing about the world. In the

[*Philosophical*] *Remarks*, Wittgenstein changes his mind on this significant question by proving that an arithmetical equation cannot be replaced by a tautology and that it therefore represents a synthetic *a priori* truth.[16]

But this is certainly taken out of the context. Although it seems clear that Wittgenstein tries to spell out what the concept of synthetic *a priori* means, there is no sign of him endorsing this Kantian concept. The reason that Wittgenstein brings in arithmetical equations into his discussion in the *Philosophical Remarks* is to show his change of mind with regard to the nature of logical forms in developing his idea of language-games, and has nothing to do with a synthetic *a priori* truth. The logical forms given along with objects in the *Tractatus* are no longer considered the given, and now Wittgenstein's position is concerned more with human activities such as calculus operated with a pencil on paper. That is why Wittgenstein eloquently says that the proposition 3+2=5 cannot be the result of an examination of concepts.[17] Gier interprets this as Wittgenstein's surprising acceptance of a synthetic *a priori*. If we have in mind the context of Wittgenstein's ongoing discussion, we can see that here Wittgenstein does not say 3+2=5 is a synthetic *a priori*. But he only means we need a sort of physical framework in which the activity of calculation takes place.

In fact, Wittgenstein's phenomenological position does not, and need not, assert synthetic *a priori*. The source of logic in the *Tractatus* is nowhere else found than in objects. All actual and possible logical forms are the logical forms of objects and their combinations. But these objects are given in immediate experience. For this reason, it is not legitimate to ask (or answer) *a priori* "the question whether I can get into a position in which I need the sign for a 27-termed relation in order to signify something."[18] This question can only be meaningful when asked (or answered) in actual experience, i.e., when our logic is empirically applied. Whether or not I need a 27-termed relation can only be shown in experience. And it has nothing to do with the logic itself that Wittgenstein discusses in the *Tractatus*. If we could know this in advance by some sort of intuition, or any other constituting activity as Husserl might want to call it, then Wittgenstein, too, might have said there is a synthetic *a priori*.[19] But, in the first place, there is no need for Wittgenstein to have any ideas like constitution. Forms are already there in materials we have in our immediate experience. The objects given us are so articulated that additional intuition or constituting activity is unnecessary. And thus, the problem of synthetic *a priori* does not take place in Wittgenstein's phenomenology.

2. INTERPRETATIONS AGAINST PHENOMENOLOGICAL INTERPRETATION

In reaction against these approaches of viewing Wittgenstein's philosophy as phenomenology, some philosophers have maintained that Wittgenstein was not a phenomenologist. One example is Harry Reeder, who strongly dismisses all claims that Wittgenstein was a phenomenologist during any period of his philosophical career.

In his paper "Wittgenstein Never Was a Phenomenologist,"[20] Harry Reeder discusses many different interpretations which view Wittgenstein's philosophy as phenomenology. As the paper title says, Reeder rejects any phenomenological interpretations given by philosophers such as Ricoeur, Spiegelberg, Gier as well as Hintikka and Hintikka. The main reasons for this rejection given by Reeder can be summed up in two ways. First, there are no phenomenological methods found in Wittgenstein's philosophy. Second, Wittgenstein never clarifies the meaning of 'phenomenology', although he uses the word. One might say that Reeder's points are absolutely right. In fact, Wittgenstein neither utilizes any sort of phenomenological methods such as phenomenological reduction and constitution, nor does he give a clear definition of the word 'phenomenology' in any of his surviving writings. However, these two points should not be used to disqualify Wittgenstein's phenomenology. Two reasons can be given.

First, as mentioned in the previous section, it is virtually impossible to prove any connection between Husserl and Wittgenstein. While we know that Husserl and Mach corresponded with each other about their philosophical concerns, it is not likely that we would discover that sort of evidence between Husserl and Wittgenstein. Thus, it would be safe to say that there is no direct connection between

Husserl and Wittgenstein. However, this point does not prove that Wittgenstein never was a phenomenologist, because there is always a possibility that one can independently develop the ideas of phenomenology. Reeder pays all his attention to Wittgenstein's unfamiliarity with Husserlian texts and methods, ignoring any possible common ground between Husserl and Wittgenstein which might be revealed by a thorough examination of their texts. In particular, Reeder lists four unique methods that he suggests as necessary conditions for phenomenology.[21] That might well be true to Husserlian phenomenology. But as Wittgenstein did not establish any philosophical relationship with Husserl, it would be unreasonable to expect some methodological details of Husserlian phenomenology from Wittgenstein.

For example, Reeder says that instead of epoché, intentionality or constitution, what we find in Wittgenstein is a vague reference to phenomena.[22] We would have to agree that Wittgenstein does not have epoché, intentionality or constitution, which is the seminal idea of Husserlian phenomenology. But I have discussed several times the reasons why no such ideas or mechanism are necessary for Wittgenstein.[23] For Wittgenstein, what Husserl wants to achieve through phenomenological reduction and constitution can be achieved without any help from special philosophical methods. In the early Wittgenstein, the notion of logic is key to understanding Wittgenstein's phenomenology. The whole logic is given with the objects of immediate experience to understand the entire structure of reality. Each and every object we are acquainted with in immediate experience holds the key to revealing the entire conceptual structure in Wittgenstein's case. There is no reason for us to suspend judgment for the transcendental turn to get at the immediately given. When Wittgenstein speaks of logical analysis, he is not talking about logic in its usual sense. Rather, he talks about conceptual analysis whose ultimate base is grounded on immediate experience. Commenting on Wittgenstein's "Some Remarks on Logical Forms," Reeder says that the atomistic and metric analysis of phenomena has nothing to do with phenomenology.[24] But as I have discussed thus far, Wittgenstein's logical analysis is indeed a phenomenological analysis in the sense that logic—the key to revealing the entire structure of reality—gets started by the objects of immediate experience. And for the same reason, Reeder should not take Felix Kaufmann's statement "phenomenology must go beyond logical analysis" as an example to disqualify Wittgenstein's logical analysis as phenomenology.[25] The lack of Husserlian methods of phenomenology should not be used for disqualifying Wittgenstein's phenomenology. Rather, we should understand that Wittgenstein did a different kind of phenomenology.

Also, Reeder's second point that Wittgenstein never clarifies what he means by his use of the word 'phenomenology' should not be used to disqualify Wittgenstein's phenomenology. The reason for this was also discussed in the previous section, where I discussed Gier's position. It was Wittgenstein's philosophical habit to give no clear definition about anything. But, most importantly, that does not mean at all that Wittgenstein only had vague ideas of some philosophical topics. It was Wittgenstein's habit that he lost all interest in a philosophical problem when he reached a conclusion after a long battle with the problem, partly because of his impatient character. The first section of the *Philosophical Remarks* is a good example that proves this point. There Wittgenstein pronounces that phenomenological language is unnecessary and impossible without giving any reason whatsoever. But the study of *Wiener Ausgabe*, the original source materials of the *Philosophical Remarks* reveals that there is philosophical dynamism displaying what Wittgenstein's problem with phenomenological language was and how he struggled with the problem to come up with the final solution. It is only one example.[26] Although Wittgenstein does not manage to put his philosophical ideas in a way in which they are stylistically and philosophically clear, we cannot say that he suffered a lack of philosophical problems and understanding. Dismissal of Wittgenstein's phenomenology just because of the lack of its clarity cannot be justified.

Another point to be noted in Reeder's critical essay is his mistaken view that Wittgenstein puts forward a form of linguistic reductionism. Reeder claims Wittgenstein reduces phenomenological problems to problems of language by rejecting phenomenological language.[27] He believes that the rejection of phenomenological language made it impossible for Wittgenstein to describe actual immanent phenomena. Since no actual description of phenomena is possible, Reeder concludes, what Wittgenstein turned to is grammar that deals with the rules of language with which language sets up its possibility of expression.[28] Wittgenstein indeed rejected phenomenological language that would do the job of immediate descriptions of immediate experience in 1929. But it does not at all mean that Wittgenstein came to believe that it is impossible to describe immediate experience in our language. What he rejected in 1929 is to view phenomenological language as a basic independent language. After this rejection, Wittgenstein clearly reinstates phenomenological language within the physicalistic language system, as a language dependent on physicalistic languages. As I have discussed earlier, his middle-period discussion of solipsism virtually coincides with Wittgenstein's re-introduction of phenomenological language within the physicalistic language system.[29] As long as it is dependent on physical framework, Wittgenstein believes, immediate description of immediate experience is allowed. Wittgenstein only puts this idea into his word 'grammar'.

Finally, Reeder comments on Hintikka and Hintikka's phenomenological reading of Wittgenstein's philosophy. Reeder attributes the Hintikkas' view to sensualistic reductionism.[30] But it is because he misunderstands the use of sense-data by Russell and Wittgenstein. Russell's 'sense-data' are similar to Lockean or Berkeleyan 'ideas', but are not quite the same. While Lockean-Berkeleyan ideas are mental, Russell's sense-data are physical, and Wittgenstein follows the Russellian view. Thus, the Hintikkas' view that identifies Tractarian objects with something analogous to Russellian sense-data cannot be attributed to sensualism or phenomenalism. Reeder says "it is the rejection of sensualism which gives phenomenology's use of the term 'experience' its wider scope."[31] But without doubt, Russell and Wittgenstein would also have rejected sensualism on the ground that sense-data are really a part of the real world.

Also, mentioning Wittgenstein's departure from Russell by the rejection of separately existing logical forms, Reeder asserts that this move cuts off the possibility for Wittgenstein to have a Husserlian kind of intuition of essences. But, to me, instead of refuting Wittgenstein's phenomenology, this is evidence that shows how Wittgenstein comes close to the phenomenological ideas independent of, as well as, different from Husserlian phenomenology. Reeder says,

> According to the Hintikkas, Wittgenstein rejects Russell's separation of logical forms from concrete objects of acquaintance as a distinction of type. If this interpretation is correct, it is certainly evidence that Wittgenstein's view in the *Tractatus* is quite distinct from Husserlian phenomenology, for Husserl always insisted that essences are of a different type from the objects of sensory experience.[32]

Indeed, Wittgenstein's view is quite distinct from Husserlian phenomenology. Thus, we may say Wittgensteinian phenomenology does not require the intuition of essences, not because there is no need for the elements of essence in Wittgenstein, but the essence in this case is given with objects of immediate experience as logical forms. This supports Hintikkas' point from which the very possibility of Wittgensteinian phenomenology gains its validity.

Among many interpretations that states there is no phenomenology found in Wittgenstein, one interesting case is to view Wittgenstein's philosophy, at least the early Wittgenstein, as phenomenalism. David Pears, who made a significant contribution in spelling out the critical relation between Russell and Wittgenstein, views the early Wittgenstein as a case of phenomenalism. In *The False Prison*, Pears states that Wittgenstein confines his concerns to the phenomenal world as opposed to the noumenal world in the Kantian sense. Rejecting the possibility that Tractarian objects are sense-data, Pears tries to find Wittgenstein's background of philosophical problems in Kant (and even in Schopenhauer).[33] He relates Wittgenstein with Kant in the sense that the former deals with a critique of the expression of thought in language in a similar spirit as the latter deals with a critique of thought. So now, the philosophy of both men becomes a philosophy aimed at drawing the limits for the legitimate uses of intellect (or of language in the case of Wittgenstein). For

Kant, the legitimacy is confined to the phenomenal world as opposed to the noumenal world. We may be able to think about the noumenal world. But our knowledge is limited to the phenomenal world only. Pears believes that Wittgenstein is on the same line of thought as Kant, by saying that the *Tractatus* "contains no speculations about a world of things-in-themselves behind the phenomena."[34] Thus, Pears's interpretation of Wittgenstein's position in the *Tractatus* is phenomenalistic instead of phenomenological. For Pears, even Wittgenstein's term 'phenomenological language' is understood as phenomenalistic language in the sense that "a phenomenological language, in Wittgenstein's usage, is always a language for describing the world as we find it; or, to put the point in a Kantian sense, it is a language for describing phenomena rather than things in themselves."[35]

But this phenomenalistic conclusion results from the view that Wittgenstein's philosophical background is found in German idealism initiated by Kant. It might be wrong to exclude Kantian elements present in Wittgenstein's early philosophy, for Wittgenstein clearly states that the aim of the *Tractatus* is to draw a limit to thought.[36] But as repeatedly mentioned, it should be understood that Wittgenstein's main concern is not about the structure or relation of what we can experience and what lies behind those phenomena. This point is aptly illustrated by Wittgenstein himself as follows:

> A phenomenon isn't a symptom of something else: it is the reality.
>
> A phenomenon isn't a symptom of something else which alone makes the proposition true or false: it itself is what verifies the proposition.[37]

From the above quote, it is clear that Wittgenstein never intended to view what is given in experience as mere phenomena or the representatives of the more fundamental things, i.e., things-in-themselves. From the given sense-data, Wittgenstein believes that the entire logical (i.e., conceptual) structure of the world can be revealed. What makes this possible is that the logical forms are given to us along with objects. Hence, Wittgenstein's main concern is directed toward what is given in immediate experience and the way we describe it.

Apparently, Pears's interpretation that Wittgenstein's phenomenological language is always a language describing the world as I find it is plausible, for indeed Wittgenstein's concern lies in the sphere of immediate experience. But this interpretation should be reconsidered when it goes on to say that it is a language describing phenomena rather than things-in-themselves in a Kantian sense. Wittgenstein's problem of immediate experience and the description of it does not come from the Kantian motive which presupposes that there is an unknowable world which lies behind the world of experience. Wittgenstein's phenomenological language might describe 'the world as I find it'. But that does not necessarily mean that the world is the phenomenal world as opposed to the noumenal world lying behind it. Wittgenstein's term 'phenomena' should be understood rather as a neutral term that does not allow us to presuppose a dichotomy of the two worlds.

In connection with Wittgenstein's term 'phenomena', what should be noted is that Tractarian objects are something very analogous to Russellian sense-data, which are not only the objects given in immediate acquaintance, but are parts of the real world. It is partly because Pears views Wittgenstein's sense-data as something other than this that makes him assert the early Wittgenstein is prone to phenomenalism. Pears says that "Wittgenstein appears not to have noticed that Russell identifies sense-data with physical events."[38] But Wittgenstein's sense-data should not be understood as more akin to Lokean or Berkeleyan ideas than Russellian sense-data. If we have to use the term sense-data in a broad sense to cover what philosophers have called sense-perception in general, then Lockean or Berkeleyan ideas are mental sense-data, which are more or less mind-dependent; whereas Russellian sense-data are physical sense-data, which are not affected by the contribution of the human mind. From Wittgenstein's remarks quoted above, we can infer that Wittgenstein's notion of sense-data is closer to the Russellian use than the Lockean or Berkeleyan use. The objects we have in phenomena, Wittgenstein says, are the reality and not a symptom of the reality. This point is strengthened by Ramsey's record. I once again quote Ramsey's record to see clearly what Wittgenstein means by Tractarian objects, and thus sense-data.

> W says [it is] nonsense to believe in anything not given in experience . . . For to be mine, to be given in experience, is [the] formal property to be a genuine entity. Other people's s.d.[=sense-data] and my s.d. i.e., those attached to my body as other people's to their bodies are logical constructions . . . . Take the criterion of what he sees to the meaning of [what] he sees. Similar I see—I see. Then I see is also connected with reality (visual appearance) [in a way] other than how his seeing is, and this is [the reason] that the world is my world.[39]

For Wittgenstein, sense-data are not only what is given in one's experience, but are also connected with reality. This, in fact, is evidence which enables us to align Wittgenstein with Russell rather than Locke and Berkeley or any philosopher who belongs to the tradition of German idealism including Kant and Schopenhauer. Hence, Wittgenstein's own remarks, coupled with Ramsey's record, refute any claim that Tractarian objects should not be identified with sense-data. Even in his Cambridge lectures, in fact, Wittgenstein is reported to have said that object and sense-datum are not two separate things.[40] For Wittgenstein, the world we live in is the world of sense-data, which are not caused by our concepts, but are themselves the sources of our concepts.[41] In this sense, sense-data are not mental but directly reflect parts of the real world without any contribution by the human mind. Therefore, Wittgenstein's whole point should not be understood as phenomenalism that confines our direct accessibility only to the phenomenal world.

The last sentence quoted above from Ramsey's record decisively shows that Wittgenstein believes that the world is constructed out of sense-data given in one's experience. Thus, what is given cannot be mental sense-data but something similar to the Russellian objects of acquaintance. Also, Wittgenstein's solipsism should be understood as sense-data based solipsism. Nevertheless, Pears views that Wittgenstein's treatment of solipsism in the *Tractatus* is concerned with the individuation of the ego, and not with the specification of the type of objects presented to the ego. He calls this view slide-peg egocentricism.[42] Similarly, on the grounds that Wittgenstein does not specify the type of objects presented to the ego, he claims

Wittgenstein's early philosophy can be viewed as a phenomenalism, i.e., slide-peg phenomenalism, which cannot specify the base-line of the phenomenal world one experiences.[43] According to Pears, Wittgenstein held this view until 1936 when he developed a criticism of the private language of sensations. To him, thus, the 1929 rejection of phenomenological language is a mere rejection of the need to translate ordinary factual language into a single basic language, which would do the complete analysis of all factual discourse.[44] Hence, for Pears, the 1929 rejection of phenomenological language cannot be a turning-point or a mark of significant transition in Wittgenstein's philosophy. He argues that Wittgenstein undertook gradual and fragmented changes rather than radical and global ones. So, in addition to the rejection of any need to have one single basic language, what Wittgenstein took up as a new philosophical concern in 1929 was the nature of immediate experience and the way we communicate about it, which is the question he had not posed before, according to Pears.[45] And this problem, which led Wittgenstein to deal with private sensations and sensation-language, is answered by the first version of the private language argument presented in the "Lecture Notes for 'Private Experience' and 'Sense Data'" in 1936.

But as I have emphasized in this study, Wittgenstein's ultimate philosophical concern throughout his career was immediate experience and how to describe it. This problem is not just confined to the status of internal objects and the method of their communication. Nor is this problem new for Wittgenstein after 1929. This is rather an overall problem that bothered Wittgenstein from the early period through the very last days of his life. When we regard Ramsey's above-quoted record as the genuine position held by Wittgenstein, it is very hard not to see that Wittgenstein dealt with something other than the nature of immediate experience and of the language that describes it; what is given in my experience is a genuine entity and the world is constructed out of objects of my experience; in this sense the world is my world; and the language describing these objects of my experience is my language. If this is Wittgenstein's early concern with immediate experience, his late period concern is found in his treatment of color concepts, as we see in *Remarks on Colour*.

Also, Pears's position that the 1929 rejection of phenomenological language is not a significant change in Wittgenstein's philosophy can be challenged. First of all, as discussed in chapter 3, section 1 above, the rejection of phenomenological language should be viewed as a major consequence from the abandonment of the completeness of truth-function theory, which also resulted in the abandonment of the independence thesis of elementary propositions. The important thing underlying this series of changes in 1929 is Wittgenstein's concern with the method of comparison between language and reality. It is relatively easy to reach a consensus among philosophers about this idea of comparison in Wittgenstein's picture theory. If we agree on this point, then everything is in order now. The truth-function theory, in fact, functioned in such a way that the very pictorial character of elementary propositions can be retained also in all other propositions, so that every proposition reflects the fact. In other words, the truth-function theory itself involves the idea of comparison between language and reality—indeed, it is not an independent theory of picture theory. The incompleteness of the truth-function theory implies the incompleteness of the comparison method. In 1929, Wittgenstein found that he was in need of modifying this comparison method, so to speak. With the incompleteness of the truth-function theory, it is inevitable to give up the independence thesis of elementary propositions. Each elementary proposition can no longer be connected to form complex propositions by means of truth-functional operations. And what this entails is serious, because it means the direct comparison between a single proposition and a single fact is no longer possible. This is the context in which Wittgenstein uses the analogy of scale and what it measures in the early middle period. Not only that, the impossibility of direct comparison necessitated Wittgenstein to bring in the idea of actual manipulation of proposition, which resulted in the idea of verification that involves the mediation of calculus-like activity taking place with symbols on paper.

Static as well as transparent comparison of language and the world is no longer possible. And a new problem arises. It is that the new method of comparison, i.e., verification, takes time. What we verify is a proposition at the time of verification, whereas what the proposition expresses is an event of the past. Thus, what a word stands for should be a temporally persisting object. In fact, Wittgenstein was looking for that kind of object in 1929. This point is vividly shown in his recently published notebooks of 1929. What we can see clearly from his notebooks is that the problem Wittgenstein struggled with in 1929 indeed involves the idea of comparison between language and reality. Wittgenstein himself mentions this point more than once:

I must be able to, in fact, compare reality with the proposition.[46]

This remark appears a few lines after Wittgenstein points out the importance of the verification idea that we must be able to bring propositions right on reality. So Wittgenstein is already struggling with regard to the comparison method. A few pages later, dated October 1929, Wittgenstein appears to worry about a significant difference found between propositions and reality, showing that he now considers the temporal nature of propositions seriously.[47] And, once again, Wittgenstein says that it is his 'main idea' that one compares propositions with reality.[48] But it seems that he could not solve the problem regarding this apparent difference in nature between propositions and reality, and is only able to make painful confessions.

The problem of the truth of proposition is escaping from me.

I am aware that the most marvelous problems lie right next to me. But I do not see them or cannot grasp them.[49]

To put Wittgenstein's problem more clearly, it is the difference in nature which exists between phenomenological states of affairs and temporal propositions, that makes comparison difficult. Wittgenstein thought he was in need of having temporally persisting objects in phenomenological states of affairs. A major candidate for these are colors. He examines the possibility, and concludes that colors in color-space cannot be temporally persisting objects which our words stand for. In fact, he comes to think that "the immediately given is in the process of constant flux"[50] that cannot be captured in temporal words. In other words, temporally persisting objects cannot be phenomenological. So now, Wittgenstein considers phenomenological language absurd,[51] and concludes that he does not hold it to be necessary any longer.[52] This concluding remark from the above-mentioned problem, which involves Wittgenstein's 'main idea', appears in the very first section of the *Philosophical Remarks*.

When we take a look at the first section of the *Philosophical Remarks*, it is natural to be puzzled by a sudden appearance of the word 'phenomenological language' and the rejection of it. Only against the background of the source material from which the *Philosophical Remarks* was rearranged, can we grasp the whole context in which Wittgenstein traveled. Indeed, the first remark of the *Philosophical Remarks* is not really the beginning that raises a series of questions, but the conclusion of a long-fought battle. The 1929 rejection of phenomenological language has its seeds in the comparison idea, and also in the nature of immediate experience, which cannot be explained by Pears's account. Wittgenstein here deals with neither phenomenalistic language, nor translation problems of our basic language. Further, what Wittgenstein means by phenomenological language can neither be a mere episode in the development of his philosophy, nor can it be properly understood without its connection to the *Tractatus*.

3. IS PHENOMENOLOGY POSSIBLE FOR WITTGENSTEIN?

We have seen that there are many different interpretations for and against Wittgenstein's phenomenology. While some claim that Wittgenstein was a phenomenologist at least during some periods of his philosophical career, others are critical of the thesis that Wittgenstein was a phenomenologist at all. Putting this controversy apart, what is for certain is that there is hard evidence that Wittgenstein himself used the world 'phenomenology' and 'phenomenological language' for a period of time. What my study has tried to show is the context in which Wittgenstein used those words, and how this usage is related to his entire philosophy. One very important strategy in this regard has been the idea that we should not be so obsessed with Husserlian phenomenological terminology and scheme. Many of those sympathetic to the phenomenological interpretation of Wittgenstein were trapped by the Husserlian scheme, and even tried unsuccessfully to establish connections between Husserl and Wittgenstein. But throughout this study, I have tried to clarify the very idea of phenomenology, and showed the idea originated from theoretical physics at the turn of century. Phenomenology is a study which attempts to understand the ultimate structure of reality from the immediately given in our experience. It not only pays a great deal of attention to immediate experience, but in so doing tries to achieve the goal of grasping reality. Thus, it cannot incorporate with the scientific method that employs hypothetical elements by which idealization and simplification are much involved. From this perspective, it will be useful and helpful to look for possible parallelism between Husserl and Wittgenstein, but will not always be desirable to identify one's phenomenology by the other.

As Husserl's phenomenology starts from the problems of the immediately given and the way we grasp reality, so does Wittgenstein's phenomenology. But the difference in the structure of the immediately given inevitably resulted in different phenomenology among the two important philosophers of this century. For Wittgenstein, this phenomenological character is most clearly present in his early period, where he tries to codify phenomenological language that would perfectly capture phenomenological objects in immediate experience. His concern in phenomenology naturally turns to immediate experience and how to describe it. The basic assumption that objects are given to us, along with their logical forms proves crucial to Wittgenstein's phenomenology. This unique view on objects, a variation of Russell's theory of acquaintance, makes it possible to ground every possible meaning in our language, and thus thought, as well as logical possibility on immediate experience. According to this view, not only does our language receive its meaning from the initial acquaintance with objects, but the entire logic reflecting the ultimate structure of the world depends on immediate experience. What makes all the complex meaning and logic of our language and thought possible depends entirely upon the logical forms of simple objects given in immediate experience. His truth-function theory only functions in the role of networking between simple logical forms and complex logical forms. No matter how complex they are does not matter, for once the logical forms of simple objects are given, the rest of the combinations takes place according to the very logical possibilities of objects from within. Thus, the logical operations by means of truth-function is a rather static job with all the possible scenarios already written. The feature that logical, or in Husserlian sense eidetic, articulation is given with the experience of objects makes much complicated apparatus like phenomenological reduction unnecessary to Wittgenstein's phenomenology. Instead, the feature that the possibility of language is given with objects naturally turns Wittgenstein's concern to phenomenological language, which transparently reflects the phenomenological world. We should be careful of this difference which does not disqualify Wittgenstein's phenomenology as phenomenology because of its lack of phenomenological reduction or other Husserlian methods, but makes Wittgenstein's phenomenology different from Husserl's.

Historical evidence that Husserl was influenced by Mach in developing and radicalizing the idea of phenomenology initiated from the field of theoretical physics can be useful here, because we can also say that Wittgenstein was much influenced by theoretical physicists' idea of phenomenology. As influences on Wittgenstein's formulation of phenomenological ideas, we can include Mach, Boltzmann, Hertz and others whose lines of thought on phenomenological physics in fact inspired Husserlian phenomenology. Therefore, disqualification of Wittgenstein's phenomenology on the ground that no elements of Husserlian phenomenology are found in Wittgenstein is mistaken. Even though Wittgenstein does not say so in the *Tractatus*, there are phenomenological elements in the book, as described above, and in that sense Wittgenstein was a phenomenologist.

Now, the problem to be discussed is what happened to his phenomenology thereafter, as we witness Wittgenstein's much determined pronouncement of the impossibility of phenomenological language in the first section of the *Philosophical Remarks*. Apparently, the impossibility of phenomenological language for Wittgenstein should make his phenomenological enterprise impossible, for his phenomenology in the *Tractatus* focuses on phenomenological language, whose function is the description of the phenomenological world, and whose possibility was based on the givenness of objects with logical forms. But Wittgenstein claims that phenomenological language is impossible, which means that his view on logical forms has changed. Initially prompted by the problem of color-incompatibility, what Wittgenstein comes to realize in the early middle period is that there can be different logical forms (and thus different symbols) according to different cases, e.g., color, sound, number, etc.[53] It is not that there is only one set of logical forms that apply to every mode of representation between language and the world. In the early system, logical forms, reflecting the possibility of the arrangement of the world, functioned as a method of projection that relates language and the world. But Wittgenstein gives up this belief in the early middle period. That resulted in subsequent abandonment of the completeness of the truth-function theory and the independence thesis of elementary propositions. And, most importantly, Wittgenstein gives up phenomenological language. All these changes revolve around the way in which we compare language and the world. At bottom, Wittgenstein tries to find what he thinks is the right method of projection and comparison, which will capture phenomenological reality. Wittgenstein's conclusion is that his earlier notion of logic does not suffice to deal with this problem. He should have to explain why we immediately know it is not blue, yellow, green or many other different colors when we identify a point in the visual field as red. Also, he should have to explain why we immediately know it is not 5 feet or 7 feet when we measure one's height and get the result of 6 feet. His solution is that language should be treated as a whole. Rather than relying on each proposition that describes each state of affairs, focus should be on the entire system of language. This change makes the Tractarian logic an insufficient tool for the later enterprise of Wittgenstein's philosophy.

And what replaces logic at this point is grammar, which is a more flexible and more inclusive notion than logic.

In later development, grammar is determined by the language-games we play. However, the replacement of logic by grammar entails a very important consequence in Wittgenstein's phenomenology. In the *Tractatus*, what makes Wittgenstein's philosophy phenomenological is the idea that logical forms can reflect the entire structure of reality. It is assumed that it is possible to reveal the essence of the world by means of logical forms. And now, what results from the abandonment of phenomenological language and the replacement of logic by grammar is that Wittgenstein can no longer retain the elements of an eidetic science that will reveal the ultimate essence of reality in his phenomenology.

Logical forms, which were once thought to be given along with objects, and by which the entire structure of the world was revealed, are now no longer available in Wittgenstein's philosophy. Wittgenstein's change of mind from viewing language as the unit of propositions to the system of language as a whole makes it unnecessary to rely on the simple in talking of meaning. Thus, since there is no longer fixed logical forms or essences in the way in which we conceptualize, phenomenology as a systematic eidetic science is impossible for Wittgenstein. One might argue that the notion of grammar in later Wittgenstein would do the job, as Wittgenstein himself once mentioned "phenomenology is grammar." Indeed, the rejection of phenomenological language and the replacement of logic by grammar do not entirely abolish phenomenology in Wittgenstein's philosophy. The phenomenology chapter in the Big Typescript is a good example showing that Wittgenstein did not entirely give up his phenomenology. Also, we have *Remarks on Colour*, in which Wittgenstein asserts that "there are phenomenological problems." Thus, it should be pointed out that Wittgenstein did not stop dealing with phenomenological problems after the 1929 rejection of phenomenological language. In fact, for Wittgenstein the phenomenological problems are the problems of immediate experience and the way we describe it.

The problem of immediate experience does not appear to have changed greatly, because we see, for example, Wittgenstein takes color—i.e., the color in immediate experience—to be one of his philosophical problems from the earliest years of his philosophy to the last. That Wittgenstein takes up color consistently in his works is easily ignored. But to one's surprise, the problem of color in immediate experience is consistently present in all periods of Wittgenstein's philosophy. The problem of color appears in the *Tractatus* with regard to logical possibility. In the *Philosophical Remarks*, Wittgenstein explicitly says that he is concerned with color as something we immediately experience. Colors also function as an important example as Wittgenstein develops the idea of the language-game. Wittgenstein discusses this problem in the *Brown Book*, *Philosophical Investigations*, and *Remarks on the Philosophy of Psychology*. And finally, perhaps there is no better example than *Remarks on Colour*, which shows how Wittgenstein consistently takes up the problem of color, for it is Wittgenstein's very last work and proves that he was still struggling with the problem of color after all those years.

As the color case shows, the problem of immediate experience has never escaped Wittgenstein's attention. It has remained as the main philosophical concern for Wittgenstein. What has changed during his philosophical development are his views on the way in which we express and describe immediate experience. Wittgenstein's rejection of phenomenological language is the rejection of one way of describing immediate experience in favor of another. And in this context, the problem of the possibility of phenomenology for Wittgenstein should be approached from the perspectives of the ways we describe immediate experience, and not of the eidetic science.

What Wittgenstein came to believe in 1929 is the primacy of physicalistic language over phenomenological language. The way we conceptualize might begin with immediate experience, and Wittgenstein appears to have never given up this point as his consistent concern with immediate experience shows. But the language that expresses immediate experience cannot be phenomenological. And that means his earlier idea of logical analysis is no longer appropriate to cover all the problems of immediate experience. In fact, logical analysis of the early period means the analysis of phenomeno-logic. Objects are given to us in immediate experience, and that is the way in which we are furnished with logic. Networking of this logic and the analysis of the way in which this networking is arranged is what Wittgenstein meant by logical analysis. But at the same time, this networking serves the purpose of describing immediate experience. It consists of ideal language in which the description of immediate experience is codified. In the middle period, however, as the possibility of this ideal language is put into doubt, Wittgenstein introduces new logic which is not static, and thus cannot simply be networked. Hence, logical analysis of phenomenological language now becomes grammatical analysis of physicalistic language characterized as a rule-governed human activity.

Viewed from the context in which Wittgenstein has traveled, having a great interest in grammar and grammatical analysis is very natural and also logical. For if Wittgenstein did not give up his concern with immediate experience, as he had given up the possibility of phenomenological language, he must have said of the way we express immediate experience. Understanding logical forms will not do the job any longer. Then the job will more or less be done by the replacement of logic, and this replacement is grammar. Of course, grammar for Wittgenstein is neither a fixed form nor is it entirely experience-based as logical forms were. Grammar can be meaningfully viewed only against the background of the use and the application of the actual language-game we play. Thus, the real core of Wittgenstein's logico-phenomenological ideas in the early period no longer exists in his later works. No fixed forms or essences can be found to formulate a systematic eidetic science. This point is also true in regard to the early Wittgenstein, regardless of the fact that he had fixed forms of logic. In the early Wittgenstein, we may analyze the way in which these logical forms operate to mirror the world. But we are not situated to analyze the way in which these logical forms are attained. Logical forms are simply given with objects. The question of why and how they are given cannot even be asked, for it essentially asks about something prior to what we can experience. It is to ask what logic is. This ultimate question, for Wittgenstein, does not come into the legitimate field of what can be said. Our language is possible by the logical forms of objects provided in experience. But language is the very means with which we think. The way in which these logical forms operate is a legitimate question we can deal with. But the essence of language can neither be expressed, nor systematically studied.

In this sense, there could not be phenomenology as a systematic science in the early Wittgenstein, even though there is phenomenology in what he said in the *Tractatus*. From the middle period on, as grammatical analysis falls short of his earlier phenomenology of the *Tractatus*, the possibility of phenomenology as systematic science is even slimmer. Therefore, all Wittgenstein could say was the analysis of grammar, by which the description of immediate experience is possible, and not to be mixed up or confused with non-phenomenological descriptions. For this reason, the very first section of the *Philosophical Remarks* implies such a point, which sets up a new direction for his philosophy from the conclusion of a long fought battle he undertook in 1928 and 1929. The first sentence of the *Philosophical Remarks* says,

> A proposition is completely logically analysed if its grammar is made completely clear: no matter what idiom it may be written or expressed in.[54]

If one is not familiar with the direction Wittgenstein has traveled as explained, this first sentence, which even precedes the rejection of phenomenological language, will be puzzling. Many possible interpretations can be made. But, in fact, this only shows Wittgenstein's change of attitudes in dealing with the same old phenomenological spirit. Earlier on, Wittgenstein thought logical analysis was nothing other than phenomenological analysis by which phenomenological states of affairs are perfectly reflected. The way in which immediate experience is expressed relied on logical analysis. Now, the way in which immediate experience is expressed does not rely on logical analysis. But if the complete clarification of grammar is achieved, then this will do the same job as logical analysis. As

Wittgenstein himself says a few lines later in the first section of the *Philosophical Remarks*, this grammatical analysis, by which we separate what is essential from what is inessential in our language, is to give "an immediate representation of immediate experience." And this grammatical analysis of recognizing what is essential and what is not indeed amounts to the same job as phenomenological language would have done. Thus, even though we have no phenomenological language, and we do not have phenomenology as a systematic science, Wittgenstein thought we could achieve the same purpose as what he intended to do in his earlier work. The phenomenology chapter of the Big Typescript also supports this idea in its opening remark:

> The investigation of rules of the use of our language, the knowledge of these rules and perspicuous representation, amounts to the thing; i.e., achieves the same thing as what we often intend to achieve by the construction of phenomenological language.[55]

The only problem Wittgenstein faces then is that he does not yet know how it would be possible to express immediate experience. That he did not develop any clear ideas for the way in which we describe immediate experience apart from the rough idea of grammar is also shown in the Big Typescript:

> The geometry of our visual space is given to us, that is, it requires no investigation in order to find hitherto hidden facts. The investigation is neither in the sense of physical nor of psychological *investigation*. However, we can say that we do not know this geometry yet. This geometry is grammar and the investigation is a grammatical investigation.[56]

This perplexity of not knowing grammar seems to be resolved soon after those years of crisis and puzzlement. In the *Blue Book*, in fact, in a lengthy discussion of solipsism, Wittgenstein shows us the ways in which phenomenological grammar and physical grammar operate in our language and conceptual scheme.[57] Phenomenology involves solipsistic grammar, which he terms 'the geometrical eye', whereas our ordinary physicalistic ways of expression involve publicly identifiable grammar, which he terms 'the physical eye'. In reality, this introduction of two different methods of identification (or grammar if we allow Wittgenstein's own terminology) amounts to one solution to the annoying problem of how to describe immediate experience. In short, his way of dealing with this problem is to allow the operation of phenomenological language within a physicalistic language system. This is followed by the private language argument in which he gives the solution to the way we express our inner sensations. In both cases, what is important is the physicalistic language framework, in which even solipsistic or phenomenological expressions, as well as the expressions of sensations, can be meaningfully situated. Thus, in a way, Wittgenstein's phenomenology after 1929 hides itself deep down in the expression of physicalistic language we use. Wittgenstein thought that it was the only way we could deal with immediate experience and the phenomenological problems. Wittgenstein could not establish a systematic science of phenomenology at all. But it was not because there were no genuine phenomenological ideas or problems for Wittgenstein, but he did not believe it was ever possible to establish the systematic science of phenomenology.

LIST OF ABBREVIATIONS

BLB *The Blue Book* in *The Blue and Brown Books: Preliminary Studies for the "Philosophical Investigations,"* ed. Rush Rhees, Oxford: Basil Blackwell, 1958.

BRB *The Brown Book* in *The Blue and Brown Books: Preliminary Studies for the "Philosophical Investigations,"* ed. Rush Rhees, Oxford: Basil Backwell, 1958.

BT The Big Typescript (TS 213)

CV *Culture and Value*, ed. G. H. von Wright in collaboration with Heikki Nyman; trans. Peter Winch, Oxford: Basil Blackwell, 1980.

LWVC *Ludwig Wittgenstein and the Vienna Circle:* Conversations recorded by Friedrich Waismann, ed. Brian McGuinness; trans. Joachim Shulte and Brian McGuinness, Oxford: Basil Blackwell, 1980.

NB *Notebooks 1914-1916*, 2nd ed., ed. G. H. von Wright and G. E. M. Anscombe; trans. G. E. M. Anscombe, Oxford: Basil Blackwell, 1979.

PE "Notes for Lectures on 'Private Experience' and 'Sense Data,'" originally published with omissions from MSS 148, 149, and 151 in *Philosophical Review* 77 (1968), ed. and trans. Rush Rhees: 274-320; reprinted, with additional materials from MSS 148, 149, and 151 in *Philosophical Occasions 1912-1951*, eds. James Klagge and Alfred Nordmann (Indianapolis: Hackett, 1993), ed. and trans. for the additional material, David Stern, pp. 202-88.

PG *Philosophical Grammar*, ed. Rush Rhees; trans. Anthony Kenny, Oxford: Basil Blackwell, 1974.

PI *Philosophical Investigations*, trans. G. E. M. Anscombe, Oxford: Basil Blackwell, 1953.

PO *Philosophical Occasions 1912-1951*, ed. James Klagge and Alfred Nordmann, Indianapolis and Cambridge: Hackett, 1993.

PR *Philosophical Remarks*, ed. Rush Rhees; trans. R. Hargreaves and R. White, Oxford: Basil Blackwell, 1975.

RC *Remarks on Colour*, ed. G. E. M. Anscombe; trans. Linda L. McAlister and Margarete Schättle, Berkeley and Los Angeles: University of California Press, 1977.

RLF "Some Remarks on Logical Form," in *Proceedings of the Aristotelian Society, Supplementary Volume*, 8 (1929): 162-71; reprinted in *Philosophical Occasions 1912-1951*, pp.29-35.

***RPP*, i** *Remarks on the Philosophy of Psychology*, vol.I, ed. G. E. M. Anscombe and G. H. von Wright; trans. G. E. M. Anscombe, Oxford: Basil Blackwell, 1980.

***RPP*, ii** *Remarks on the Philosophy of Psychology*, vol.II, ed. G. H. von Wright and Heikki Nyman; trans. C. G. Luckhardt and M. A. E. Aue, Oxford: Basil Blackwell, 1980.

TLP *Tractatus Logico-Philosophicus*, trans. D. F. Pears and B. F. McGuinness, London: Routledge & Kegan Paul, 1961.

WA *Wiener Ausgabe*, ed. Michael Nedo from MSS 105-108, Vienna: Springer-Verlag, 1994.

NOTES

PREFACE

[1] *Wittgenstein's Lectures, Cambridge, 1930-1932*, ed. Desmond Lee (Oxford: Basil Blackwell, 1980), p. 82.
[2] *RC*, I. sec. 53; III. sec. 248.
[3] Merrill B. Hintikka and Jaakko Hintikka, *Investigating Wittgenstein* (Oxford: Basil Blackwell, 1986).

CHAPTER I

[1] They have been catalogued as MSS 105-108 by G.H. von Wright, and published recently as *Wiener Ausgabe* (Vienna: Springer Verlag, 1994-95).
[2] *Philosophical Remarks* consists of the materials Wittgenstein prepared for the renewal of his research grant at Cambridge. It is Wittgenstein's own selection and arrangement of passages from MSS 105-107 and the first half of 108. For the full account of its composition, see 'Editor's Note' in *PR*, pp. 347-51.
[3] The Big Typescript (TS 213) is believed to be the only work Wittgenstein produced in a conventional book format divided by chapters and sections. It even contains a table of contents, which suggests that he at one time intended to publish it. But only parts of the whole typescript have been published in the *Philosophical Grammar*. For a detailed account regarding the publication of the *Philosophical Grammar*, see Anthony Kenny's "From the Big Typescript to the *Philosophical Grammar*" in *The Legacy of Wittgenstein* (Oxford: Basil Blackwell, 1984).
[4] M.O'C. Drury, "Conversations with Wittgenstein," in *Ludwig Wittgenstein, Personal Recollections*, ed. Rush Rhees (Totowa, NJ: Rowman & Littlefield, 1981), p. 131.
[5] Rush Rhees, "The *Tractatus*: Seeds of Some Misunderstandings," *Philosophical Review* 72 (1963): 217.
[6] *PR*, sec. 218.
[7] See *LWVC*, pp. 67-8.
[8] There Wittgenstein says: "I do not now have phenomenological language, or 'primary language' as I used to call it, in mind as my goal. I no longer hold it to be necessary, All that is possible and necessary is to separate what is essential from what is inessential in *our* language" (*PR*, sec. 1).
[9] After the rejection of phenomenological language in section 1 of the *Philosophical Remarks*, Wittgenstein explicitly says that our language is physicalistic in the following remarks: "Language itself belongs to the second

system. If I describe a language, I am essentially describing something that belongs to physics" (*PR*, sec. 68). For a detailed discussion of Wittgenstein's shift in language paradigms, see chapter 3, section 1 below.

[10] See *WA*, i., p. 23; *PR*, secs. 1, 11, 50, 53, 57, 71, 75; *LWVC*, p. 45.

[11] *PR*, sec. 11.

[12] *PR*, sec. 57.

[13] Also, we should not confuse his shift of language paradigms with the trend in philosophy that attempts to reduce all psychological (and other) concepts to naturalistic terms. From his use of the term 'physicalistic language', there is no evidence that Wittgenstein is advancing or endorsing the idea of reductionism.

[14] Rudolf Carnap, *Logical Structure of the World and Pseudoproblems in Philosophy*, 2nd ed., [*Der Lohische Aufbau der Welt* (Berlin: Weltkreis-Verlag, 1928)] trans. Rolf George (London: Routledge & Kegan Paul, 1967), sec. 67, p. 108.

[15] See Carnap, *The Unity of Science*, ["Die physicalische Sprach als Universalsprache der Wissenschaft," *Erkenntnis* (1932): 432-65)], trans. Max Black (London: Kegan Paul, 1934), pp. 44-5.

[16] It is rather more likely that Wittgenstein influenced Carnap's use of the term 'primary language' and the idea of physicalism. See Merrill B. Hintikka and Jaakko Hintikka, *Investigating Wittgenstein* (Oxford: Basil Blackwell, 1986), pp. 146-7; Jaakko Hintikka, "Ludwig's Apple Tree: On the Philosophical Relations between Wittgenstein and the Vienna Circle," in *Scientific Philosophy: Origins and Developments* ed. F. Stadler (Dordrecht: Kluwer Academic Publishers, 1993), p. 33.

[17] Frank Ramsey, *Foundations of Mathematics and Other Logical Essays* (London: Kegan Paul, 1931), p. 212.

[18] Herbert Spiegelberg, *The Phenomenological Movement*, 3rd ed. (The Hague: Martinus Nijhoff, 1982). See Introduction, pp. 1-24.

[19] *Ibid.*, p. 9.

[20] Ernst Mach, "On the Principle of Comparison in Physics"(1894), in *Popular Scientific Lectures*, trans. Thomas J. McCormack (Chicago: The Open Court, 1898), p. 250.

[21] Ludwig Boltzmann, "On the Development of the Methods of Theoretical Physics in Recent Times," in *Theoretical Physics and Philosophical Problems*, ed. Brian McGuinness (Dordrecht: D. Reidel, 1974), p. 95.

[22] *Ibid.*, p. 95.

[23] See *Ernst Mach—A Deeper Look*, ed. John Blackmore (Dordrecht: Kluwer Academic Publishers, 1992), chap. 6, pp. 127-50.

[24] "The Meaning and Limits of Exact Science," in *Scientific Autobiography and Other Papers* (New York: Philosophical Library, 1949), pp. 80-120.

[25] Albert Einstein, "Physics and Reality," in *Ideas and Opinions* (New York: Crown Publishers, 1954), pp. 302-3.

[26] See Edmund Husserl, "Amsterdamer Vorträge," in *Phänomenologische Psychologie: Husserliana* Vol. 9 (The Hague: Martinus Nijhoff, 1962), pp. 302-3.
[27] Spiegelberg, *The Phenomenological Movement*, 2nd ed., p. 9.
[28] See Keiichi Noé, "Mach's Relativism versus Apriorism and Mechanistic World View," in *Ernst Mach—A Deeper Look*, ed. John Blackmore, *op. cit.*, pp. 237-8. Interestingly, it is reported that the Dutch philosopher Evert W. Beth even claimed that Husserl stole basic ideas of phenomenology from Mach. See Henk Visser, "Mach, Utrecht, and Dutch Philosophy," in *Ernst Mach—A Deeper Look*, ed. Blackmore, *op. cit.*, p. 429.
[29] See *LWVC*, p. 67-8.
[30] *CV*, p. 19.
[31] G.H. von Wright, *Wittgenstein* (Minneapolis: University of Minnesota Press, 1981), p. 213.
[32] Brian McGuinness, *Wittgenstein: A Life* (Berkeley: University of California Press, 1988), p. 38. Boltzmann's (as well as Hertz's) influence on Wittgenstein is studied also in Andrew D. Wilson, "Hertz, Boltzmann and Wittgenstein Reconsidered," *Studies in the History and Philosophy of Science* 20 (1989): 245-63. Cf. Allan Janik, "How Did Hertz Influence Wittgenstein's Philosophical Development?" *Grazer Philosophische Studien* 49 (1994/95): 19-47.
[33] *PR*, sec. 1.
[34] Boltzmann, "On the Development of the Methods of Theoretical Physics," *op. cit.*, p. 96. For Boltzmann's relation to phenomenology, metaphysics and related philosophical issues, see also John Blackmore, "Introduction," in *Ludwig Boltzmann: His Later Life and Philosophy, 1900-1906*, ed. John Blackmore (Dordrecht: Kluwer Academic Publishers, 1995), especially pp. 27-33, with further references.
[35] Boltzmann, *op. cit.*, p. 97.
[36] Boltzmann, "On Statistical Mechanics," *op. cit.*, pp. 155-6.
[37] See Jaakko Hintikka, "The Idea of Phenomenology in Wittgenstein and Husserl," in *Ludwig Wittgenstein: Half-Truths and One-and-a-Half-Truths* (Dordrecht: Kluwer Academic Publishers, 1996).
[38] The paper that provoked Wittgenstein's outrage is "Die physikalische Sprache als Univesalsprache der Wissenschaft," published in *Erkenntnis* in 1932. For a detailed story about this incident and the relation between Wittgenstein and the Vienna Circle, see Jaakko Hintikka, "Ludwig's Apple Tree," in *Scientific Philosophy: Origins and Development*, ed. F. Stadler (Dordrecht: Kluwer, 1993), pp. 27-46.; Thomas E. Uebel, "Physicalism in Wittgenstein and the Vienna Circle," in *Physics, Philosophy and the Scientific Community*, eds. K. Gavroglu et al. (Dordrecht: Kluwer Academic Publishers, 1995), pp. 327-56.
[39] Rudolf Carnap, *Logical Structure of the World and Pseudoproblems in Philosophy* (London: Routledge & Kegan Paul, 1967), sec. 64, p. 101.
[40] *Ibid.*, sec. 93, p. 145. See also sec. 152, pp. 233-4.

[41] *Ibid.*, sec. 65, pp. 103-6.
[42] A general exposition of solipsism in relation to phenomenology is discussed in chapter 2, section 4. Also the similarity of Carnap's methodological solipsism to Wittgenstein's middle-period treatment of solipsism is discussed in chapter 3, section 2.
[43] G. E. Moore, "Refutation of Idealism," in *Philosophical Studies* (London: Routledge & Kegan Paul, 1922), p. 27. Moore's point of color as independent object of consciousness is also found in Wittgenstein's treatment of color. When Wittgenstein deals with color in the *Philosophical Remarks* and the *Remarks on Colour*, it is the color we immediately experience, and not the one physics or psychology would deal with. (Cf. Chapter 2, footnote 3 below.) Also, the general philosophical influence of Moore on Wittgenstein has been emphasized by J. N. Findlay. See his *Wittgenstein: A Critique* (London: Routledge & Kegan Paul, 1984), pp. 68-73.
[44] See Bertrand Russell, "The Relation of Sense-Data to Physics," in *Mysticism and Logic and Other Essays* (London: George Allen & Unwin, 1963), pp. 111-3, and "Knowledge by Acquaintance and Knowledge by Description," *op. cit.*, p. 167.
[45] Bertrand Russell, *The Problems of Philosophy* (Oxford: Oxford University Press, 1969), p. 13.
[46] See Russell, "The Relation of Sense-Data to Physics," p. 113.
[47] *Ibid.*
[48] For a detailed discussion of this point, see chapter 2, section 1 below.
[49] Edmund Husserl, *Ideas of Phenomenology*, trans, William Alston and George Nakhnikian (The Hague: Martinus Nijhoff, 1964), p. 3.
[50] *Ibid.*, p. 16.
[51] See Jaakko Hintikka, "The Phenomenological Dimension," in *The Cambridge Companion to Husserl*, eds. Barry Smith and David W. Smith (Cambridge: Cambridge University Press, 1995), pp. 78-105.
[52] Husserl, *Idea of Phenomenology*, p. 10.
[53] *Ibid.*, pp. 48-9.
[54] Husserl, *Cartesian Meditations*, trans. Dorion Cairns (The Hague: Martinus Nijhoff, 1973), sec. 8, pp. 20-1.
[55] See Russell, "The Relation of Sense-Data to Physics," p. 109.

CHAPTER II

[1] Russell, "Knowledge by Acquaintance and Knowledge by Description," in *Mysticism and Logic and Other Essays* (London: George Allen & Unwin, 1963), p. 152.
[2] Russell, *The Problems of Philosophy* (Oxford: Oxford University Press, 1969), p. 46.

[3] As in the case of Moore's objects of consciousness, Russell's notion of color as something we are directly aware of is further developed by Wittgenstein's treatment of color, which is given in our immediate experience. So even the earliest notion of color-incompatibility (*TLP*, 6.3751) should be understood in this context.

[4] Russell, "On the Nature of Acquaintance," in *Logic and Knowledge*, ed. Robert C. Marsh (London: George Allen & Unwin, 1956), p. 167.

[5] *NB,* entry dated 16 June 1915, p. 61.

[6] Russell, *Theory of Knowledge: The 1913 Manuscript—The Collected Papers of Bertrand Russell Vol. 7*, ed. Elizabeth R. Eames in collaboration with Kenneth Blackwell (London: George Allen & Unwin, 1984). Also see David Pears, "The Relation between Wittgenstein's Picture Theory of Propositions and Russell's Theory of Judgment," in *Wittgenstein: Sources and Perspectives*, ed. C. G. Luckhardt (Ithaca: Cornell University. Press, 1979) and Hintikka and Hintikka, *Investigating Wittgenstein* (Oxford: Basil Blackwell, 1986), chap. 3.

[7] Russell, *Theory of Knowledge*, p. 97.

[8] *Ibid.*, p. 99.

[9] *Ibid.*

[10] *Ibid.*, p. 132.

[11] Wittgenstein, *Letters to Russell, Keynes and Moore*, ed. G. H. von Wright (Ithaca: Cornell University. Press, 1974), p. 23.

[12] David Pears, "The Relation between Wittgenstein's Picture Theory of Propositions and Russell's Theory of Judgment," p. 206.

[13] See *NB*, entries dated 22 August 1914, p. 2; 2 September 1914, p. 2; 3 September 1914, p. 2; 13 October 1914, p. 11; and *TLP* 5.473.

[14] *NB*, dated 29 October 1914, p. 21.

[15] See Hintikka and Hintikka, *Investigating Wittgenstein*, pp. 56-7.

[16] *NB*, entry dated 1 November 1914, p. 23 .

[17] *NB*, entry dated 3 November 1914, p. 24.

[18] *NB*, entry dated 4 November 1914, p. 26.

[19] *NB*, entry dated 18 October 1914, p. 15.

[20] Hintikka and Hintikka, *Investigating Wittgenstein*, pp. 53-4.

[21] Russell, "On the Nature of Acquaintance," p. 167.

[22] See Editor's Preface in *LWVC*, p. 22.

[23] *LWVC*, p. 250.

[24] *LWVC*, p. 249.

[25] *LWVC,* p. 254.

[26] *BRB*, pp. 171-2.

[27] *Wittgenstein's Lectures, Cambridge, 1930-1932*, ed. Desmond Lee (Oxford: Basil Blackwell, 1980), p. 81.

[28] *Ibid*, p. 82.

[29] Pittsburgh Ramsey archives, item #004-21-02. Quoted in Hintikka and Hintikka, *Investigating Wittgenstein*, p. 77.
[30] Husserl, *Ideas Pertaining to a Pure Phenomenology and to a Phenomenological Philosophy: First Book*, trans. F. Kersten (Dordrecht: Kluwer, 1982), sec. 50, p. 113.
[31] *Ibid*, sec. 97, p. 239.
[32] See Hintikka, "Husserl: The Phenomenological Dimension," in *The Cambridge Companion to Husserl*, eds. Barry Smith and David W. Smith (Cambridge: Cambridge University Press, 1995), p. 97.
[33] *Ibid.*
[34] In fact, in his *Logical Investigations*, Husserl describes something very similar to Russell's logical forms when he discusses categorial intuition. For example, Husserl says, "what intuitively corresponds to the words 'and' and 'or', to 'both' or 'either', is not anything, as we rather roughly put it above, that can be grasped with one's hands, or apprehended with some sense, as it can also not really be represented in an image, e.g., in painting. I can paint A and I can paint B, and I can paint them both on the same canvas: I cannot, however, paint the *both*, nor paint the A *and* B." (Husserl, *Logical Investigations*, vol. 2, trans. J. N. Findlay, New York: Humanities Press, 1970, p. 798) The words 'and', 'or', 'both' and 'either' are examples of Russell's molecular complex which constitutes logical forms along with atomic complex. See Russell's *Theory of Knowledge*, pp. 80, 99. But a more interesting fact is that Husserl sometimes uses the word 'logical forms' or 'categorial forms' in the discussion of categorial intuition. See *Logical Investigations*, pp. 818-9.
[35] Hintikka, "Idea of Phenomenology in Wittgenstein and Husserl," in *Ludwig Wittgenstein: Half-Truths and One-and-a-Half-Truths* (Dordrecht: Kluwer, 1996).
[36] Husserl, *Ideas Pertaining to a Pure Phenomenology and to a Phenomenological Philosophy*, sec. 27, p. 51.
[37] Husserl, *Cartesian Meditations*, p. 2.
[38] *Ibid.*, pp. 35-7.
[39] See Dagfinn Føllesdal, "An Introduction to Phenomenology for Analytic Philosophers," in *Contemporary Philosophy in Scandinavia*, eds. Raymond E. Olson and Anthony M. Paul (Baltimore: The Johns Hopkins Press, 1972), p. 426.
[40] Husserl, *Cartesian Meditations*, p. 67.
[41] *Ibid.*, p. 65.
[42] *Ibid.*, p. 67.
[43] See Russell, "Relation of Sense-Data to Physics," in *Mysticism and Logic*, p. 112.
[44] Russell, *Problems of Philosophy*, p. 19.
[45] Russell, "Knowledge by Acquaintance and Knowledge by Description," in *Mysticism and Logic*, p. 162.
[46] See section 5 of this chapter, footnote 51 below.

[47] See Jaakko Hintikka, "On Wittgenstein's 'Solipsism'," *Mind* 67 (1958): 88-91.
[48] Cf. see *PR*, sec. 67. Wittgenstein says, "Suppose I had such a good memory that I could remember all my sense impressions. In that case, there would, *prima facie*, be nothing to prevent me from describing them. This would be a biography."
[49] *NB*, 7 August 1916, p. 80.
[50] *NB*, 11 August 1916, p. 80.
[51] See Russell, "Knowledge by Acquaintance and Knowledge by Description," pp. 153-4. In this paper (written in 1910), Russell argues that "it would seem necessary, therefore, either to suppose that I am acquainted with myself, and that 'I', therefore, requires no definition, being merely the proper name of a certain object, or to find some other analysis of self-consciousness." However, in his 1914 paper "On the Nature of Acquaintance," Russell excludes 'I' from the true proper names. "It follows that the word 'I', as commonly employed, must stand for a description; it cannot be a true proper name in the logical sense, since true proper names can only be conferred on objects with which we are acquainted." See Russell, "On the Nature of Acquaintance," p. 164.
[52] See Jaakko Hintikka, "Wittgenstein and the Problem of Phenomenology," in *Language, Knowledge, and Intentionality: Acta Philosophica Fennica*, Vol. 49, eds. Leila Haaparauta, Martin Kusch, and Ilkka Niiniluoto (Helsinki: Philosophical Society of Finland, 1990), p. 17.
[53] See *PR*, sec. 71. "It could, e.g., be practical under certain circumstances to give proper names to my hands and to those of other people, so that you wouldn't have to mention their relation to somebody when talking about them, since that relation isn't essential to the hands themselves; and the usual way of speaking could create the impression that its relation to its owner was something belonging to the essence of the hand itself."

Also sec. 74. "Even the word 'visual space' is unsuitable for our purpose, since it contains an allusion to a sense organ which is as inessential to the space as it is to a book that it belongs to a particular person."
[54] See Anthony Kenny, "The First Person," in *The Legacy of Wittgenstein* (Oxford: Basil Blackwell, 1984), pp. 77-87.
[55] Hintikka and Hintikka, *Investigating Wittgenstein*, pp. 106-9.
[56] *NB*, p. 98. Also cf. *TLP*, 4.024, 5.5151.
[57] See Jaakko Hintikka, "An Anatomy of Wittgenstein's Picture Theory," in *Artifacts, Representations and Social Practice*, eds. C. C. Gould and R. S. Cohen (Dordrecht: Kluwer Academic Publishers, 1994), p. 228.
[58] "Some Remarks on Logical Forms" was originally published in *Proceedings of the Aristotelian Society, Supplementary Volume*, vol. 9, and reprinted in *PO*, pp. 29-35.

[59] The Tractarian scheme of truth possibilities is presented in *TLP* 5.101. Wittgenstein's example with color-incompatibility is shown in "Some Remarks on Logical Form," in *PO*, pp. 33-5.
[60] *PO*, pp. 29-30.
[61] The propositions of the *Tractatus* in question read as follows:

3.323 In everyday language it very frequently happens that the same word has different modes of signification—and so belongs to different symbols—or that two words that have different modes of signification are employed in propositions in what is superficially the same way.

3.324 In this way the most fundamental confusions are easily produced (the whole of philosophy is full of them).

3.325 In order to avoid such errors we must make use of a sign-language that excludes them by not using the same sign for different symbols and by not using in a superficially similar way signs that have different modes of signification: that is to say, a sign-language that is governed by *logical* grammar—by logical syntax.

[62] *PO*, p. 30.
[63] *PO*, p. 35.
[64] See *PR*, sec. 78.
[65] *PR*, sec. 78.
[66] *PR*, sec. 79.
[67] *PR*, sec. 82.
[68] *PR*, Appendix 2, p. 317. Also, *LWVC*, p. 63.
[69] Cf. Wittgenstein himself elaborates this point in *LWVC*, pp. 73-4.
[70] See *PR*, sec. 3. Wittgenstein states, "The words 'Colour', 'Sound', 'Number' etc. could appear in the chapter headings of our grammar." Cf. also sec. 81.
[71] *PR*, sec. 83.

CHAPTER III

[1] The first entry of MS 105 is dated February 2, 1929 (See *WA*, i., p. 3); in May of 1930 Wittgenstein showed the typescript version of the *Philosophical Remarks* to Russell. See Editor's note in *PR*, p. 347.
[2] Jaakko Hintikka, "Die Wende der Philosophie: Wittgenstein's New Logic of 1928" in *Philosophy of Law, Politics, and Society: Proceedings of the 12th International Wittgenstein Symposium*, eds. Elisabeth Leinfellner and Rudolf Haller (Vienna: Hölder-Pichler-Tempski, 1987), p. 381.
[3] *RLF*, reprinted in *PO*, p. 31.
[4] *LWVC*, p. 47.
[5] *PR*, sec. 45

[6] *WA*,ii., p. 89. The German original is: "Man muß den Satz mit der Wirklichkeit *zur Deckung bringen* können/auf die Wirklichkeit auflegen können/." (All translations are my own unless otherwise noted.)
[7] See *WA*,ii., p. 91.
[8] 'Time' here means physical time (i.e., information time) as opposed to memory time. For more detailed treatment on Wittgenstein's notion of time, see Jaakko Hintikka "Wittgenstein on Being and Time," in *Ludwig Wittgenstein: Half-Truths and One-and-a-Half-Truths* (Dordrecht: Kluwer Academic Publishers, 1996). About Wittgenstein's phenomenology of time, see section 3 of this chapter below.
[9] See *WA*,ii., p. 92.
[10] *LWVC*, p. 45.
[11] *PR*, sec. 57.
[12] See Jaakko Hintikka, "Wittgenstein and the Problem of Phenomenology," in *Language, Knowledge, and Intentionality: Acta Philosophica Fennica*, vol. 49, eds. Leila Haaparauta, Martin Kusch, and Ilkka Niiniluoto (Helsinki: Philosophical Society of Finland, 1990), p. 20.
[13] *BLB*, p. 51.
[14] Hintikka, "Wittgenstein and the Problem of Phenomenology," p. 23.
[15] *BLB*, p. 46.
[16] *BLB*, p. 48.
[17] *BLB*, p. 58.
[18] See Hintikka, "Witgenstein and the Problem of Phenomenology," pp. 25-30. Cf. For recent studies that pay attention to Wittgenstein's 'geometrical eye', see David Pears, "The Ego and the Eye: Wittgenstein's Use of an Analogy," *Grazer Philosophische Studien* 44 (1993): 59-68. Also see Michel ter Hark, *Beyond the Inner and the Outer* (Dordrecht: Kluwer Academic Publishers, 1990), especially pp. 82-93.
[19] *BLB,* p. 64.
[20] *BLB*, p. 64.
[21] See *PR*, sec. 210.
[22] See *PR*, secs. 206, 208.
[23] *BRB*, p. 109.
[24] *BLB,* p. 59.
[25] *BLB*, p. 57.
[26] *BLB*, p. 61.
[27] *BLB*, p. 67.
[28] *BLB*, p. 71.
[29] *PE*, reprinted in *PO*, p. 255.
[30] *PE*, reprinted in *PO*, p. 258.
[31] *BLB*, p. 71.
[32] *BLB*, p. 72.
[33] *PR*, sec. 71.

[34] *PR*. sec. 74.
[35] *PE*, reprinted in *PO*, p. 258.
[36] Carnap, *The Logical Structure of the World* (London: Routledge & Kegan Paul, 1967), sec. 65,p. 104.
[37] *Ibid*., sec. 16, p. 29.
[38] See chapter 1, section 2 above.
[39] G. E. Moore, "Wittgenstein's Lectures in 1930-33," originally published in *Mind* 63-64 (1954-1955), reprinted in *PO*, p.110.
[40] See Jaakko Hintikka, "Wittgenstein on Being and Time," in *Ludwig Wittgenstein: Half-Truths and One-and-a-Half-Truths* (Dordrecht: Kluwer Academic Publishers, 1986).
[41] Russell, *Our Knowledge of the External World* (Chicago: Open Court, 1914), pp. 121-2.
[42] There is an interpretation of Russell's theory of acquaintance that cannot be made quite consistent with Russell's idea of time-construction based on memory. For example, J. O. Urmson says that "it is impossible to use proper names of things in their absence," and that "a logically proper name can be given only to an object of acquaintance while one is actually acquainted with it; one cannot use demonstrative symbol to name a thing which is not present." (See Urmson, *Philosophical Analysis* [Oxford: The Clarendon Press, 1956], pp. 86, 134.) But David Pears criticizes Urmson's view by saying that "a person is said to have knowledge of a thing by acquaintance if he has come across it in his experience and remembers it, or, alternatively, if he is experiencing it at the moment." (See Pears, *Bertrand Russell and the British Tradition in Philosophy* [New York: Random House], 1967, p. 71.) As will be shown, I think Pears's view is more reasonable as we witness that Russell, in fact, regards memory as an immediate source of knowledge.
[43] Russell, *The Problems of Philosophy*, p. 48.
[44] Russell, "Knowledge by Acquaintance and Knowledge by Descriptions," in *Mysticism and Logic*, p. 152.
[45] Russell, "The Relation of Sense-Data to Physics," in *Mysticism and Logic*, p. 123.
[46] See *PR*, sec. 67.
[47] Moore, "Wittgenstein's Lectures in 1930-33," in *PO*, p. 110.
[48] *Ibid*., p. 111.
[49] Russell, *The Problems of Philosophy*, p. 22.
[50] *Ibid*.
[51] Moore presents in his lecture notes the different views between Russell and Wittgenstein as follows: "It looks to me as if, for the moment, Wittgenstein was confusing the two entirely different propositions, (1) 'It is logically possible that nothing exists except the present experience' which Russell may be said to imply, and (2) 'It is logically *im*possible that anything should exist except the present

experience', which he certainly does not imply." (*PO*, p. 111) Certainly, Russell would have believed (1) because, although temporal solipsism is logically not impossible, there is no reason to believe it, as there is no reason to believe experiential solipsism. Thus, Russell proposed to construct time from memory. On the other hand, what Wittgenstein saw in Russell is that if we rely on memory alone, we inevitably fall into temporal solipsism like (2).

[52] See chapter 2, section 5 above.

[53] At times, however, Wittgenstein relates temporal solipsism to experiential one. For example, he says: "If someone says, only the *present experience* has reality, then the word 'present' must be redundant here, as the word 'I' is in other context. . ." (*PR*, sec. 54, p. 85)

[54] Moore, "Wittgenstein's Lectures," in *PO*, p. 112.

[55] *BLB*, p. 67.

[56] *BRB*, p. 108.

[57] *BRB*, p. 108-9.

[58] *PR*, sec. 67.

[59] See Husserl, *On the Phenomenology of the Consciousness of Internal Time (1893-1917)*, trans. John Barnett Brough (Dordrecht: Kluwer Academic Publishers, 1991), pp. 31, 66.

[60] See *ibid.*, pp. 42-3.

[61] One interesting thing here is that Husserl's ideas of time is also very similar to that of William James. James in the chapters entitled "The Perception of Time" and "Memory" in *The Principles of Psychology* elaborates his ideas on 'specious present', 'primary memory', and 'secondary memory'. Husserl's ideas of 'now-point', 'primary memory' and 'secondary memory' seem to have their origin in James's corresponding terms as Husserl acknowledges his debt to James as early as the time of *Logical Investigations*. The fact that Wittgenstein has read James's book and even refers to him in the *Philosophical Investigations* and other late writings indeed shows how the respective interests of Husserl and Wittgenstein come close. J. N. Findlay seems to notice this respective interest between Husserl and Wittgenstein. But while he credits James and Husserl for what they have done regarding time and memory, Findlay discredits Wittgenstein's contribution based on the contents of the *Brown Book*. Regarding James's influence on Husserl, see Herbert Spiegelberg, *The Phenomenological Movement*, 3rd ed. (The Hague: Martinus Nijhoff, 1982), pp. 100-4. For James's accounts of time and memory, see *The Principles of Psychology*, 2 vols (New York: Henry Holt & Company, 1890), especially vol.1, pp. 630-55. And for Findlay's comments on this matter, see his *Wittgenstein: A Critique* (London: Routledge & Kegan Paul, 1984), pp. 158-60.

[62] For Wittgenstein's later criticism of this view, see section 5 of this chapter.

[63] Jaakko Hintikka, "Rules, Games and Experiences: Wittgenstein's Discussion of Rule-Following in the Light of His Development," in *Revue internationale de philosophie*, 43 (1989): 285.
[64] According to Hintikka, Wittgenstein's rule-following discussion (*PI*, secs. 143-242) includes his effort to show his change of mind regarding this matter. For later Wittgenstein, experience is not the essential element involved in rules and rule-following. What becomes the most fundamental element is the language-game idea. Without the basis of language-games, no question of rule and rule-following can be meaningfully asked.
[65] These remarks appear in MS 116, sec. 218, and quoted here from Hintikka, "Rules, Games and Experience," p. 285.
[66] *BLB*, p. 1.
[67] *BLB*, p. 2.
[68] *BLB*, p. 37.
[69] The development of this idea seems to have a long history. Briefly, Wittgenstein frequently used the word 'calculus' in *LWVC, PR,* and *PG*, roughly from 1929 to early 1933. But from the *Blue Book* on, Wittgenstein hardly used the word after his self-criticism that language in practice is not used by strict rules as in calculus. It is true that our language is a rule-governed activity, but not the strict system as calculus governed by mathematical rules. So Wittgenstein opted to see our language as more like children's outdoor ball game. See *BLB*, p. 25.
[70] "I shall in the future again and again draw your attention to what I shall call language games. These are ways of using signs simpler than those in which we use the signs of our highly complicated everyday language." *BLB*, p. 17.
[71] *PG*, p. 244.
[72] This is the reason why there cannot be private language, even if we could utilize private rules arbitrarily. See the following section for a detailed discussion of private language.
[73] Hintikka and Hintikka, *Investigating Wittgenstein*, pp. 187-8.
[74] *Ibid.*, p. 189.
[75] *BLB*, pp.12-3.
[76] See *PI*, secs. 156-159.
[77] See *PI*, secs. 197, 198.
[78] Hintikka, "Rules, Games and Experiences," p. 295.
[79] A similar remark appears in *Remarks on the Philosophy of Psychology* as follows: "The primitive language-game we originally learned needs no justification, and false attempts at justification, which force themselves on us, need to be rejected." (*RPP*, ii., sec. 453)
[80] *PI*, sec. 655.
[81] See Hintikka and Hintikka, *Investigating Wittgenstein*, chap. 10. especially pp. 253-5.

[82] That Wittgenstein does not mean the private 'use' of language when he talks about private language is shown in section 243 of the *Philosophical Investigations*. He says "But could we also imagine a language in which a person could write down or give vocal expression to his inner experiences—his feelings, moods, and the rest—for his private use?—Well, can't we do so in our ordinary language?—*But that is not what I mean*." (Emphasis added.)

Wittgenstein is certainly not interested in the matter of empirical fact in this regard. Instead, the point he makes in his discussion of private language is a conceptual one.

[83] Jaakko Hintikka, "Wittgenstein on Knowledge and Skepticism," forthcoming.

[84] Hintikka and Hintikka call this framework of spontaneous pain-behavior as physiognomic framework. Their view is that the connection between sensations and sensation-behavior is a logical one by means of which private experiences are correlated to the language-game. See Hintikka and Hintikka, *Investigating Wittgenstein*, p. 258.

[85] *PE*, reprinted in *PO*, p. 258.

[86] Wittgenstein began to write what is believed to be the prototype of the rule-following discussion (*PI*, secs. 143-242) around 1936 with MS 115. But the source materials for sections 198-242, including the explicit analogy of the primacy of language-games to 'customs' constituted by 'sign-posts' in sections 198 and 199, were mostly written in 1944-45. The exceptions are sections 214, 215, and 239 only. Moreover, what constitutes the so-called private language argument (from secs. 242 to 315) were also written mostly in 1944-45. Hence, we can say that Wittgenstein's idea of the primacy of language-games was not fully developed until 1944. About the source materials of the *Philosophical Investigations*, see André Maury, "Sources of the Remarks in Wittgenstein's *Philosophical Investigations*," *Synthese* 98 (1994): 349-78.

CHAPTER IV

[1] In *TLP*, 2.0131 Wittgenstein says "A speck in the visual field, though it need not be red, must have some colour: it is, so to speak, surrounded by colour-space."

[2] In *TLP*, 2.0251 Wittgenstein says "Space, time, and colour (being coloured) are forms of objects."

[3] This point is in sharp distinction to Wittgenstein's use of 'colors' as examples of simple objects. See Desmond Lee's *Wittgenstein's Lectures: Cambridge 1930-1932*, p. 120.

[4] Rush Rhees, "The *Tractatus* : Seeds of Some Misunderstandings" *Philosophical Review* 72 (1963): 217.

[5] *PR*, sec. 218.

[6] See *PR*, sec. 218.

[7] In a letter (19 January, 1950) to von Wright, Wittgenstein writes "The last two weeks I read a great deal in Goethe's 'Fabenlehre'. It's partly boring and repelling, but in some ways also *very* instructive and philosophically interesting." See "Letters from Ludwig Wittgenstein to Georg Henrik von Wright," in *PO*, p. 475.
[8] See *PR*, sec. 221.
[9] See *PR*, sec. 219.
[10] See *PR*, sec. 220.
[11] See Moore's "Wittgenstein's Lectures in 1930-33," in *PO*, p. 108.
[12] See *PR*, sec. 221.
[13] *LWVC*, p. 67.
[14] Cf. Hintikka and Hintikka suggest a solution to the problem which arises from the color-incompatibility issue that misleads people to see the impossibility of having two colors at the same time as a physical impossibility. They propose to see "This patch is green" and "This patch is red" not as the expressions of subject-predicate forms, but as functions that will map points in visual space into color-space. So instead of translating the two sentences into *'G(a)'* and *'R(a)'*, Hintikka and Hintikka propose to use functional analysis which translates the above sentences into *'C(a)=r'* and *'C(a)=g"* which enable us to see that there can only be one value for *a*, and thus that having two values, say green and red, at the same time is a logical impossibility. For a detailed discussion, see Hintikka and Hintikka, *Investigating Wittgenstein*, chap. 5, especially secs. 3-5.
[15] *LWVC*, p. 241.
[16] *BLB*, p. 56.
[17] Wittgenstein says, "We are not, however, regarding the language games which we describe as incomplete parts of a language, but as languages complete in themselves, as complete system of human communication." *BRB*, p. 81.
[18] See *BRB*, p. 85.
[19] See chapter 3, section 3 above.
[20] *PE*, reprinted in *PO*, p. 220.
[21] *PE*, reprinted in *PO*, p. 220.
[22] Also see *PO*, p. 219.
[23] *BRB*, p. 87.
[24] *BRB*, pp. 136-7.
[25] See *RPP*, ii., secs. 312, 313, and 315.
[26] *PE*, reprinted in *PO*, pp. 234-5.
[27] *PE*, reprinted in *PO*, p. 235.
[28] *RC*, III, sec. 71.
[29] *RC*, III, sec. 72.
[30] See Jaakko Hintikka, "The Cartesian *Cogito*, Epistemic Logic and Neuroscience," *Synthese* 83 (1990): 146-7. Also, Lucia M. Vaina, "'What' and

'Where' in the Human Visual System: Two Hierarchies of Visual Modules," *Synthese* 83 (1990): 49-91.
[31] *Wittgenstein's Lectures, Cambridge, 1930-1932* ed. Desmond Lee (Oxford: Basil Blackwell, 1980), p. 82.
[32] *Ibid.*
[33] See Hintikka and Hintikka, *Investigating Wittgenstein*, pp. 298-30.
[34] See *PI*, secs. 273, 274.
[35] See *PI*, sec. 276.
[36] See *PI*, sec. 280.
[37] Even before he developed the idea of language-games, Wittgenstein wrestled with this problem of expressing immediate experience. One example is his observation that, although the visual space differs from the physical space, we cannot successfully express what we see in our visual field without using the words we usually use to describe the physical space. For instance, there can be no measurement in the visual field. But still, we use the words like 'exact', 'precise', etc., which seem to be more appropriate words for the description of the physical space where measurement can be made. For a detailed discussion, see chapter 3, section 2 above.
[38] *RC*, I, sec. 22. Also see *RC*, III, sec. 188. This remark reminds one of Wittgenstein's middle-period treatment of the same problem. In the *Philosophical Remarks*, Wittgenstein says: "What I need is a psychological or rather phenomenological colour theory, not a physical and equally not a physiological one." (*PR*, sec. 218)
[39] *RC*, II, sec. 16.
[40] *RC*, I, sec. 39; III, sec. 221.
[41] *PR*, sec. 39.
[42] See *RC*, I, secs. 10, 11, 14, 21; III, secs. 27, 30, 52. Also *RPP*, ii., secs. 421, 422, 423, 428.
[43] *RPP*, ii., sec. 422. A similar example appears in *RC*, III, sec. 30.
[44] *RC*, I, sec. 6.
[45] *RC*, I, sec. 19.
[46] See *RC*, I, sec. 37 and III, sec. 156.
[47] See *RC*, I, sec. 40.
[48] *RC*, III, sec. 76.
[49] See *RC*, I, sec. 51 and III, sec. 229.
[50] *RC*, III, sec. 234.
[51] *RC*, I, sec. 63 and III, sec. 117. Also cf. III. secs. 273, 276.
[52] *RC*, III, sec. 59.
[53] *RC*, III, sec. 35. Also see I, sec. 3.
[54] David Katz, *The World of Colour*, trans. by R.B. MacLeod (London: Kegan Paul, 1935).
[55] *Ibid*, p. 2.

[56] See *ibid.*, p. 3.
[57] *RC*, I, sec. 77 and III, sec. 120.
[58] *RC*, III, sec. 319.
[59] See *BLB*, pp. 13-4; *BRB*, pp. 85-8; *PI*, secs. 48, 49.
[60] Wittgenstein says there: "For naming and describing do not stand on the same level: naming is a preparation for description. Naming is so far not a move in the language-game—any more than putting a piece in its place on the board is a move in chess. We may say: *nothing* has so far been done, when a thing has been named. It has not even *got* a name except in the language-game." (*PI*, sec. 49)
[61] See chapter 3, section 5 above.
[62] *RC*, I, sec. 53.
[63] *RC*, III, sec. 248.

CHAPTER V

[1] *BRB*, p. 175.
[2] *BRB*, p. 172.
[3] See Jaakko Hitikka and Merrill B. Hintikka, "Ludwig Looks at the Necker Cube" in *Acta Philosophica Fennica*, vol. 38, edited by Ghita Holmström and Andrew Jones (Helsinki: The Philosophical Society of Finland, 1985), p. 43.
[4] According to the *OED*, the definition of 'aspect' includes "the appearance presented by an object to the eye." See *Oxford English Dictionary*, 2nd ed. (Oxford: Clarendon Press, 1989), vol. 1, p. 692. Indeed, this definition serves the best purpose for Wittgenstein's phenomenology, for his phenomenological object is not physical object, but the object given in our experience, i.e., presented to the eye.
[5] *PG*, p. 167. Emphasis added.
[6] *PG*, p. 166. Emphasis added.
[7] See *PG*, pp. 174 and 444.
[8] *LWVC*, pp. 256-7.
[9] *LWVC*, p. 257.
[10] *LW*, i, sec. 634. See also *PI*, p. 206 where Wittgenstein says "The same tone of voice expresses the dawning of aspect."
[11] *LW*, i, sec. 677.
[12] TS 213, p. 442. The German original is: "Eine Kirchentonart verstehen, heißt nicht, sich an die Tonfolge gewöhnen, in dem Sinne, in dem ich mich an einen Geruch gewöhnen kann und ihn nach einiger Zeit nicht mehr unangenehm empfinde. Sondern es heißt, etwas Neues hören, was ich früher noch nicht gehört habe, etwa in der Art—ja ganz analog—wie es wäre, 10 Striche ||||||||||, die ich früher nur als 2 mals 5 Striche habe sehen können, plötzlich als ein charakteristisches Ganzes sehen zu können. Oder die Zeichnung eines Würfels, die ich nur als flaches Ornament habe sehen können, auf einmal räumlich zu

sehen." This passage also appears in *PR*, sec. 224 and *WA*, ii, p. 221. The translation above is from the *PR*.
[13] Language-game (8) is presented in *PI*, sec. 8, in which 'there' and 'this' among others are used in connection with a pointing gesture.
[14] *PI*, sec. 38.
[15] *BRB*, p. 171.
[16] *TLP*, 5.5423.
[17] Cf. *PI*, p. 208.
[18] *LW*, i, sec. 588.
[19] See *RPP*, ii, sec. 451; *PI*, p. 207.
[20] *PI*, p. 209.
[21] *PI*, p. 208.
[22] *RPP*, i, sec. 862.
[23] *LW*, ii, sec. 549; *PI*, p. 197.
[24] *LW*, i, sec. 571; cf. *PI*, p. 198.

CHAPTER VI

[1] C. A. van Peursen, "Edmund Husserl and Ludwig Wittgenstein," *Philosophy and Phenomenological Research* 20 (1959): 181-97.
[2] For example, the German edition of the *Philosophical Remarks* was published in 1964, and the English edition in 1975.
[3] Paul Ricoeur, "Husserl and Wittgenstein on Language," in *Phenomenology and Existentialism*, eds. E. N. Lee and M. Mandelbaum (London: Johns Hopkins Press, 1967), pp. 207-17.
[4] *Ibid.*, p. 211.
[5] Cf. Martin Kusch, *Language as Calculus and Language as Universal Medium* (Dordrecht: Kluwer Academic Publishers, 1989).
[6] The fruitful outcome of von Wright's valuable research on the *Nachlass* materials has a long publication history over the last two and a half decades. It was originally published as "The Wittgenstein Papers," *Philosophical Review* 78 (1969): 483-503. It reappeared in a revised and expanded version in von Wright's book *Wittgenstein* (Oxford: Basil Blackwell, 1982), pp. 35-62. And once again, further revised and expanded version of it was published in an Appendix to the *Philosophical Occasions, 1912-1951* (Indianapolis: Hackett, 1993), pp. 480-506. Also, about the plans for the publication of the entire *Nachlass* materials, see Jaakko Hintikka, "An Impatient Man and His Papers" *Synthese* 87 (1991): 183-201.
[7] Herbert Spiegelberg, "The Puzzles of Wittgenstein's Phänomenologie (1929-?)(with Supplement 1979)," in *The Context of the Phenomenological Movement*, ed. Herbert Spiegelberg (The Hague: Martinus Nijhoff, 1981), p. 227.

[8] See Herbert Spiegelberg, *The Phenomenological Movement* (The Hague: Martinus Nijhoff, 1982), 3rd. ed., p. 23, footnote 11.
[9] Nicholas Gier, "Wittgenstein's Phenomenology Revisited," *Philosophy Today* (Fall 1990): 274-5.
[10] *Ibid.*, 275.
[11] *PI*, sec. 46.
[12] *Wittgenstein's Lectures: Cambridge, 1930-1932*, p. 82.
[13] Nicholas Gier, *Wittgenstein and Phenomenology* (Albany: State University of New York Press, 1981), p. 104.
[14] The quoted remark from the *Philosophical Investigations* was written in 1937. About the sources of the remarks in the book, see André Maury's "Sources of the Remarks in Wittgenstein's *Philosophical Investigations*" *Synthese* 98 (1994): 349-78. And the quoted remark from the *Blue Book* is believed to be written in 1933-34. See the editor's Preface to *The Blue and Brown Book.*
[15] *PR*, secs. 57, 218. Quoted in Gier's *Wittgenstein and Phenomenology*, p. 105.
[16] Gier, *Wittgenstein and Phenomenology*, p. 158. Also see Gier, "Wittgenstein's Phenomenology Revisited," p. 275.
[17] *PR*, sec. 107.
[18] See *TLP*, 5.5541 and 5.5542.
[19] *A priori* element is inevitably implicit in Husserl's system, for the phenomenological reduction which brackets out almost every presupposed elements from consciousness makes it impossible for us to dispense with the contribution of the subject part.
[20] Harry Reeder, "Wittgenstein Never Was a Phenomenologist," *Journal of the British Society for Phenomenology* 20 (1989): 257-76.
[21] *Ibid.*, p. 258.
[22] See *ibid.*, pp. 263-4.
[23] See chapter 2, section 3 and chapter 6, section 1 above.
[24] *Ibid.*, p. 261.
[25] *Ibid.*, p. 274, footnote 22.
[26] For a detailed discussion of Wittgenstein's impatient character and the way it was present in his philosophy, see Jaakko Hintikka, "An Impatient Man and His Papers," *Synthese* 87 (1991): 183-201.
[27] Reeder, "Wittgenstein Never Was a Phenomenologist," p. 268.
[28] See *ibid.*
[29] See chapter 3, section 2 above.
[30] *Ibid.*, p. 269.
[31] *Ibid.*, p. 270.
[32] *Ibid.*
[33] See David Pears, *The False Prison*, vol.1 (Oxford: Oxford University Press, 1987), chaps. 1 and 2.
[34] *Ibid.*, p. 4.

[35] *Ibid.*, p. 98.
[36] *TLP*, Preface, p. 3.
[37] *PR*, sec. 225; *WA*, ii., p. 118.
[38] Pears, *The False Prison*, vol. 2 (Oxford: Oxford University Press, 1988), p. 278, footnote 16.
[39] Pittsburgh Ramsey archives, item #004-21-02. Quoted in Hintikka and Hintikka, *Investigating Wittgenstein*, p. 77.
[40] *Wittgenstein's Lectures, Cambridge 1930-1932*, ed. Desmond Lee (Oxford: Basil Blackwell, 1980),p. 109.
[41] *Ibid.*, pp. 81-2.
[42] Pears, *The False Prison*, vol. 2, p. 233.
[43] *Ibid.*, p. 235.
[44] See *ibid.*, chap.11, and also "Hintikka's Wittgenstein," forthcoming.
[45] Pears, "Hintikka's Wittgenstein," p. 17.
[46] *WA*, ii., p. 89. The German original is: "Ich muß die Wirklichkeit ja tatsächlich mit dem Satz vergleichen können."
[47] See *WA*, ii., p. 91.
[48] *WA*, ii., p. 91(dated 7 October 1929).
[49] *WA*, ii., p. 91 (dated 9 October 1929). The German original is: "Das Problem der Wahrheit eines Satzes entschlüpft mir." "Ich bin mir bewußt daß die herrlichsten Probleme in meiner nächsten Nähe liegen. Aber ich sehe sie nicht oder kann sie nicht fassen."
[50] *WA*, ii., p. 92 (dated 11 October 1929). German original: "Das Unmittelbare ist in ständigem Fluß begriffen." (my translation)
[51] *WA*, ii., p. 102 (dated 22 October 1929).
[52] *WA*, ii., p. 108 (dated 25 October 1929).
[53] See *PR*, sec. 3.
[54] *PR*, sec. 1.
[55] *BT*, p. 437. The German original is: "Die Untersuchung der Regeln des Gebrauchs unserer Sprache, die Erkenntnis dieser Regeln und übersichtliche Darstellung, läuft auf das hunaus, d.h., leistet dasselbe, was man oft durch die Konstruktion einer phänomenologischen Sprache leisten will."
[56] *BT*, p. 444. The German original is: "Die Geomtrie unseres Gesichtsraumes ist uns gegeben, d.h., es bedarf keiner Untersuchung bis jetzt verborgener Tatsachen, um sie zu finden. Die Untersuchung is keine, im Sinn einer physikalischen oder psychologischen *Untersuchung*. Und doch kann man sagen, wir kennen diese Geometrie noch nicht. Diese Geomtrie ist Grammatik und die Untersuchung eine grammatische Untersuchung."
[57] See chapter 3, section 2 above for details.

BIBLIOGRAPHY

1. Works by Wittgenstein

Tractatus Logico-Philosophicus. Translated by D. F. Pears and B. F. McGuinness. London: Routledge & Kegan Paul, 1961. (Originally published as *Logisch-Philosophische Abhandlung* in 1921.)

Philosophical Investigations. Translated by G. E. M. Anscombe. Oxford: Basil Blackwell, 1953.

Remarks on the Foundations of Mathematics, Edited by G. H. von Wright, Rush Rhees, and G. E. M. Anscombe; translated by G. E. M. Anscombe. Oxford: Basil Blackwell, 1956.

The Blue and Brown Books: Preliminary Studies for the "Philosophical Investigations." Edited by Rush Rhees. Oxford: Basil Blackwell, 1958.

Notebooks 1914-1916. 2nd ed., Edited by G. H. von Wright and G. E. M. Anscombe; translated by G. E. M. Anscombe. Oxford: Basil Blackwell, 1979. (1st ed., 1961).

Zettel. Edited by G. E. M. Anscombe and G. H. von Wright; translated by G. E. M. Anscombe. Oxford: Basil Blackwell, 1967.

On Certainty. Edited by G. E. M. Anscombe and G. H. von Wright; translated by Denis Paul and G. E. M. Anscombe. Oxford: Basil Blackwell, 1969.

Philosophical Grammar. Edited by Rush Rhees; translated by Anthony Kenny. Oxford: Basil Blackwell, 1974.

Philosophical Remarks. Edited by Rush Rhees; translated by R. Hargreaves and R. White. Oxford: Basil Blackwell, 1975.

Remarks on Colour. Edited by G. E. M. Anscombe; translated by Linda L. McAlister and Margarete Schättle. Berkeley and Los Angeles: University of California Press, 1977.

Culture and Value. Edited by G. H. von Wright in collaboration with Heikki Nyman; translated by Peter Winch. Oxford: Basil Blackwell, 1980.

Remarks on the Philosophy of Psychology, vol.I. Edited by G. E. M. Anscombe and G. H. von Wright; translated by G. E. M. Anscombe. Oxford: Basil Blackwell, 1980.

Remarks on the Philosophy of Psychology, vol.II. Edited by G. H. von Wright and Heikki Nyman; translated by C. G. Luckhardt and M. A. E. Aue. Oxford: Basil Blackwell, 1980.

Last Writings on the Philosophy of Psychology, vol.I. Edited by G. H. von Wright and Heikki Nyman; translated by C. G. Luckhardt and M. A. E. Aue. Oxford: Basil Blackwell, 1982.

Last Writings on the Philosophy of Psychology, vol.II. Edited by G. H. von Wright and Heikki Nyman; translated by C. G. Luckhardt and M. A. E. Aue. Oxford: Basil Blackwell, 1992.

Philosophical Occasions 1912-1951. Edited by James Klagge and Alfred Nordmann. Indianapolis and Cambridge: Hackett Publishing Company, 1993.

Wiener Ausgabe, 3 vols. Edited by Michael Nedo from MSS 105-108. Vienna: Springer-Verlag, 1994-95.

"Phänomenologie" chapter from the Big Typescript (TS 213). Unpublished.

2. Wittgenstein's Letters, Lecture Notes and Conversations Recorded by Others.

Letters to Russell, Keynes and Moore. Edited by G. H. von Wright. Oxford: Basil Blackwell, 1974.

Wittgenstein's Lectures, Cambridge, 1932-1935. Edited by Alice Ambrose. Oxford: Basil Blackwell, 1979.

Ludwig Wittgenstein and the Vienna Circle: Conversations recorded by Friedrich Waismann. Edited by Brian McGuinness; translated by Joachim Shulte and Brian McGuinness. Oxford: Basil Blackwell, 1980.

Wittgenstein's Lectures, Cambridge, 1930-1932. Edited by Desmond Lee. Oxford: Basil Blackwell, 1980.

3. Other Materials

Arrington, Robert. "Wittgenstein and Phenomenology." *Philosophy Today* 22 (1978): 287-300.

Bell, David. *Husserl*. London: Routledge, 1990.

Black, Max. *A Companion to Wittgenstein's 'Tractatus'*. Ithaca: Cornell University Press, 1964.

Blackmore, John (ed.). *Ernst Mach—A Deeper Look*. Dordrecht: Kluwer Academic Publishers, 1992.

Boltzmann, Ludwig. *Theoretical Physics and Philosophical Problems*. Edited by Brian McGuinness; translated by Paul Foulkes. Dordrecht: D. Reidel, 1974.

Brand, Gerd. *The Essential Wittgenstein*. Translated by Robert E. Innis. Oxford: Basil Blackwell, 1979.

Brenner, William. "Wittgenstein's Color-Grammar." *Southern Journal of Philosophy* 20 (1982): 289-98.

Budd, Malcolm. *Wittgenstein's Philosophy of Psychology*. London: Routledge. 1989.

Carnap, Rudolf. *The Logical Structure of the World and Pseudoproblems in Philosophy*, 2nd ed. Translated by Rolf A. George. London: Routledge &

Kegan Paul, 1967. (Originally published as *Der Logische Aufbau der Welt*, Berlin: Weltkreis-Verlag, 1928.)

Carnap, Rudolf. *The Unity of Science*. Translated by Max Black. London: Kegan Paul, 1934. (Originally published as "Die physicalische Sprach als Universalsprache der Wissenschaft.," *Erkenntnis* 1932: 432-65.)

Carnap, Rudolf. "Intellectual Autobiography." In *The Philosophy of Rudolf Carnap*. Edited by Schilpp, P. A. La Salle (Ill.): Open Court. 1963.

Casati, Roberto. "Notes on Phenomenology and Visual Space." In *Wittgenstein: Mind and Language*. Edited by R. Egidi. Dordrecht: Kluwer Academic Publishers, 1995.

Cunningham, Suzanne. *Language and the Phenomenological Reductions of Edmund Husserl*. The Hague: Martinus Nijhoff, 1976.

Dreyfus, Hubert L (ed.). *Husserl, Intentionality, and Cognitive Science*. Cambridge: The MIT Press, 1982.

Drury, M. O'C. "Conversations with Wittgenstein." In *Ludwig Wittgenstein: Personal Recollection*. Edited by Rush Rhees. Totowa, N.J.: Rowman and Littlefield, 1981.

Einstein, Albert. "Physics and Reality." In *Ideas and Opinions*, New York: Crown Publishers, 1954.

Fay, Thomas A. "Hermeneutical Phenomenology of Language in the Later Heidegger and Wittgenstein." *Dialogos* 27 (1992): 19-35.

Findlay, J. N. *Wittgenstein: A Critique*. London: Routledge & Kegan Paul, 1984.

Føllesdal, Dagfinn. "Husserl's notion of noema." *Journal of Philosophy* 66 (1969):
680-7.

Føllesdal, Dagfinn. "An Introduction to Phenomenology for Analytic Philosophers" in *Contemporary Philosophy in Scandinavia*. Edited by. Raymond E. Olson and Anthony M. Paul, Baltimore: The Johns Hopkins Press, 1972.

Frege, Gottlob. *The Foundations of Arithmetic*. Translated by J. L. Austin. New York: Harper Torchbook, 1960.

Frege, Gottlob. *Translations from the Philosophical Writings of Gottlob Frege*. Edited and translated by Peter Geach and Max Black. Oxford: Basil Blackwell. 1960.

Friedman, Michael. "Carnap's *Aufbau* Reconsidered." *Nous* 21 (1987): 521-45.

Gier, Nicholas F. *Wittgenstein and Phenomenology*. Albany: State University of New York Press, 1981.

Gier, Nicholas F. "Wittgenstein's Phenomenology Revisited." *Philosophy Today*, (Fall 1990): 273-88.

Gier, Nicholas. "Never Say 'Never': A Response to Reeder's 'Wittgenstein Never Was a Phenomenologist." *Journal of the British Society for Phenomenology* 22 (1991): 80-3.

Hallett, Garth. *A Companion to Wittgenstein's "Philosophical Investigations."* Ithaca: Cornell University Press, 1977.

Hallett, Garth. "The Genesis of Wittgenstein's Later Philosophy in His Failure as a Phenomenologist." *Philosophy and Theology* (1991): 297-311.

Haller, Rudolf. "Wittgenstein and Physicalism." *Crítica* 21 (1989): 17-32.

Hardin, C. L. *Color for Philosophers*. Indianapolis: Hackett, 1988.

Hark ter, Michel. *Beyond the Inner and the Outer*. Dordrecht: Kluwer Academic Publishers, 1990.

Hems, John M. "Husserl and/or Wittgenstein." *International Philosophical Quarterly* 8 (1968): 547-78.

Hertz, Heinrich. *The Principles of Mechanics*. Translated by D. E. Jones and J. T. Walley. New York: Dover Publications, 1956.

Hintikka, Jaakko. "On Wittgenstein's 'Solipsism'." *Mind* 67 (1958): 88-91.

Hintikka, Jaakko. "Die Wende der Philosophie: Wittgenstein's New Logic of 1928" in *Philosophy of Law, Politics, and Society: Proceedings of the 12th International Wittgenstein Symposium*. Edited by Elisabeth Leinfellner and Rudolf Haller. Vienna: Hölder-Pichler-Tempski, 1987. Reprinted in Jaakko Hintikka. *Ludwig Wittgenstein: Half-Truths and One-and-a-Half-Truths*. Dordrecht: Kluwer Academic Publishers. 1996.

Hintikka, Jaakko. "Rules, Games and Experiences: Wittgenstein's Discussion of Rule-Following in the Light of His Development." *Revue internationale de philosophie* 43 (1989): 279-97. Reprinted in Jaakko Hintikka. *Ludwig Wittgenstein: Half-Truths and One-and-a-Half-Truths*. Dordrecht: Kluwer Academic Publishers. 1996.

Hintikka, Jaakko. "The Cartesian *cogito*, Epistemic Logic and Neuroscience." *Synthese* 83 (1990): 133-57.

Hintikka, Jaakko. "Wittgenstein as a Philosopher of Immediate Experience." In *Wittgenstein—Towards a Re-Evaluation: Proceedings of the 14th International Wittgenstein Symposium*. Edited by Rudolf Haller and Johannes Brandl. Vienna: Hölder-Pichler-Tempski, 1990. Reprinted in Jaakko Hintikka. *Ludwig Wittgenstein: Half-Truths and One-and-a-Half-Truths*. Dordrecht: Kluwer Academic Publishers. 1996.

Hintikka, Jaakko. "Wittgenstein and the Problem of Phenomenology." In *Language, Knowledge, and Intentionality: Acta Philosophica Fennica*, vol.49. Edited by Leila Haaparuta, Martin Kusch, and Ilkka Niiniluoto. Helsinki: Philosophical Society of Finland, 1990. Reprinted in Jaakko Hintikka. *Ludwig Wittgenstein: Half-Truths and One-and-a-Half-Truths*. Dordrecht: Kluwer Academic Publishers. 1996.

Hintikka, Jaakko. "An Impatient Man and His Papers," *Synthese* 87 (1991): 183-201. Reprinted in Jaakko Hintikka. *Ludwig Wittgenstein: Half-Truths and One-and-a-Half-Truths*. Dordrecht: Kluwer Academic Publishers. 1996.

Hintikka, Jaakko. "Ludwig's Apple Tree: On the Philosophical Relation between Wittgenstein and the Vienna Circle." In *Scientific Philosophy: Origin and Developments*. Edited by F. Stadler. Dordrecht: Kluwer Academic Publishers, 1993. Reprinted in Jaakko Hintikka. *Ludwig Wittgenstein: Half-Truths and One-and-a-Half-Truths*. Dordrecht: Kluwer Academic Publishers. 1996.

Hintikka, Jaakko. "An Anatomy of Wittgenstein's Picture Theory." In *Artifacts, Representations and Social Practice*. Edited by C. C. Gould and R. S. Cohen. Dordrecht: Kluwer Academic Publishers, 1994. Reprinted in Jaakko Hintikka. *Ludwig Wittgenstein: Half-Truths and One-and-a-Half-Truths*. Dordrecht: Kluwer Academic Publishers. 1996.

Hintikka, Jaakko. "Husserl: The Phenomenological Dimension." In *The Cambridge Companion to Husserl*. Edited by David W. Smith and Barry Smith. Cambridge: Cambridge University Press. 1995. Reprinted in Jaakko Hintikka. *Ludwig Wittgenstein: Half-Truths and One-and-a-Half-Truths*. Dordrecht: Kluwer Academic Publishers. 1996.

Hintikka, Jaakko. "Wittgenstein on Being and Time." In *Ludwig Wittgenstein: Half-Truths and One-and-a-Half-Truths*. Dordrecht: Kluwer Academic Publishers. 1996.

Hintikka, Jaakko. "The Idea of Phenomenology in Wittgenstein and Husserl." In *Ludwig Wittgenstein: Half-Truths and One-and-a-Half-Truths*. Dordrecht: Kluwer Academic Publishers. 1996.

Hintikka, Jaakko. *Ludwig Wittgenstein: Half-Truths and One-and-a-Half-Truths*. Dordrecht: Kluwer Academic Publishers. 1996.

Hintikka, Merrill B. and Jaakko Hintikka. *Investigating Wittgenstein*. Oxford: Basil Blackwell, 1986.

Hintikka, Merrill B. and Jaakko Hintikka. "Wittgenstein's *annus mirabilis* : 1929." In *The Tasks of Contemporary Philosophy: Proceedings of the 10th International Wittgenstein Symposium*. Edited by Werner Leinfellner and Franz M. Wuketits. Vienna: Hölder-Pichler-Tempsky,1986. Reprinted in Jaakko Hintikka. *Ludwig Wittgenstein: Half-Truths and One-and-a-Half-Truths*. Dordrecht: Kluwer Academic Publishers. 1996.

Husserl, Edmund. "Amsterdamer Vorträge." In *Phänomenologische Psychologie: Husserliana* Vol.9. The Hague: Martinus Nijhoff, 1962.

Husserl, Edmund. *Ideas of Phenomenology*. Translated by William Alston and George Nakhnikian. The Hague: Martinus Nijhoff, 1964.

Husserl, Edmund. *Logical Investigations*. 2 vols. Translated by J. N. Findlay. New York: Humanities Press, 1970.

Husserl, Edmund. *The Crisis of European Sciences and Transcendental Phenomenology*. Translated by David Carr. Evanston: Northwestern University Press, 1970.

Husserl, Edmund. *Cartesian Meditations*. Translated by Dorion Cairns. The Hague: Martinus Nijhoff, 1973.

Husserl, Edmund. "Phenomenology", Edmund Husserl's Article for the *Encyclopaedia Britannica* (1927). In *Husserl, Shorter Works*. Edited by Peter McCormick and Frederick A. Ellison. Indiana: University of Notre Dame Press, 1981.

Husserl, Edmund. *Ideas Pertaining to a Pure Phenomenology and to a Phenomenological Philosophy: First Book*. Translated by F. Kersten. Dordrecht: Kluwer Academic Publishers, 1982.

Husserl, Edmund. *On The Phenomenology of the Consciousness of Internal Time (1893-1917)*. Translated by John Barnett Brough. Dordrecht: Kluwer Academic, 1991.

James, William. *The Principles of Psychology*, 2 vols. New York: Henry Holt &Company, 1890.

Janik, Allan. "How Did Hertz Influence Wittgenstein's Philosophical Development?." *Grazer Philosophische Studien* 49 (1994/95): 19-47.

Katz, David. *The World of Colour*. Translated by R. B. MacLead. London: Kegan Paul, 1935.

Kenny, Anthony. "The First Person." In *The Legacy of Wittgenstein*. Oxford: Basil Blackwell, 1984.

Kenny, Anthony. "From the Big Typescript to the *Philosophical Grammar*." In *The Legacy of Wittgenstein*. Oxford: Basil Blackwell, 1984.

Kusch, Martin. *Language as Calculus and Language as Universal Medium*. Dordrecht: Kluwer Academic Publishers, 1989.

Levy, Paul. *Moore: G. E. Moore and the Cambridge Apostles*. Oxford: Oxford University Press, 1981.

Mach, Ernst. "On Principle of Comparison in Physics." In *Popular Scientific Lectures*. Translated by Thomas J. McCormack. Chicago: The Open Court, 1898.

Malcolm, Norman. *Ludwig Wittgenstein, A Memoir*. London: Oxford University Press, 1958.

Malcolm, Norman. *Nothing Is Hidden*. Oxford: Basil Blackwell, 1986.

Marion, Mathieu, "Wittgenstein in Transition: Phenomenology, Hypotheses, and Assertions." Forthcoming.

Maury, André. "Sources of the Remarks in Wittgenstein's Philosophical Investigations." *Synthese* 98 (1994): 349-78.

Mayer, Verena. "Carnap und Husserl." In *Wissenschaft und Subjktivität*. Edited by David Bell and Wilhelm Vossenkuhl. Berlin: Akademie Verlag, 1992.

McGuinness, Brian. *Wittgenstein: A Life*. Berkeley and Los Angeles: University of California Press, 1988.

Merleau-Ponty, Maurice. *Phenomenology of Perception*. Translated by Colin Smith. London: Routledge & Kegan Paul, 1962.

Mohanty, J. N. *Phenomenology and Ontology*. The Hague: Martinus Nijhoff, 1970.

Mohanty, J. N. *Transcendental Phenomenology*. Oxford: Basil Blackwell, 1989.

Monk, Ray. *Ludwig Wittgenstein: The Duty of Genius*. New York: The Free Press, 1990.

Moore, G. E. "The Refutation of Idealism." In *Philosophical Studies*. London: Routledge & Kegan Paul, 1922.

Mounce, H. O. *Wittgenstein's Tractatus: An Introduction*. Oxford: Basil Blackwell, 1981.

Munson, Thomas N. "Wittgenstein's Phenomenology." *Philosophy and Phenomenological Research* 23 (1962): 37-50.

Noë, Robert Alva. "Wittgenstein, Phenomenology and What It Makes Sense to Say." *Philosophy and Phenomenological Research* 54 (1994): 1-42.

Passmore, John. *A Hundred Years of Philosophy*, 2nd ed. Harmondsworth: Penguin Books, 1968.

Pears, David. *Bertrand Russell and the British Tradition in Philosophy*. New York: Random House, 1967.

Pears, David. "The Relation between Wittgenstein's Picture Theory of Propositions and Russell's Theories of Judgment." In *Wittgenstein: Sources and Perspectives*. Edited by C.G. Luckhardt. Ithaca: Cornell University Press. 1979.

Pears, David. *Ludwig Wittgenstein*. Cambridge: Harvard University Press, 1986.

Pears, David. *The False Prison*, 2 vols. Oxford: Clarendon Press. 1987-1988.

Pears, David. "The Structure of the Private Language Argument." *Revue internationale de philosophie* 43 (1989): 264-78.

Pears, David. "The Ego and The Eye: Wittgenstein's Use of an Analogy." *Grazer Philosophische Studien* 44 (1993): 59-68.

Pears, David. "Hintikka's Interpretation of Wittgenstein's Treatment of Sensation-Language." *Grazer Philosophische Studien* 49 (1994/95): 1-18.

Pears, David. "Hintikka's Wittgenstein." Forthcoming.

Planck, Max. "The Meaning and Limits of Exact Science." In *Scientific Autobiography and Other Papers*. Translated by Frank Gaynor. New York: Philosophical Library, 1949.

Ramsey, Frank. *Foundations of Mathematics and Other Logical Essays*. London: Kegan Paul, 1931.

Reeder, Harry. *Language and Experience: Descriptions of Living Language in Husserl and Wittgenstein*. Lanham: University Press of America, 1984.

Reeder, Harry. "Wittgenstein Never Was a Phenomenologist." *Journal of the British Society for Phenomenology* 20 (1989): 257-76.

Reeder, Harry. "Never Say 'Never Say "Never"': A Reply to Nicholas Gier." *Journal of the British Society for Phenomenology* 22 (1991): 97-8.

Rhees, Rush. "The *Tractatus:* Seeds of Some Misunderstandings. *"Philosophical Review* 72 (1963): 213-20.

Rhees, Rush (ed.). *Ludwig Wittgenstein: Personal Recollection*. Totowa, N.J.: Rowman and Littlefield, 1981.

Ricoeur, Paul. "Husserl and Wittgenstein on Language." In *Phenomenology and Existentialism*. Edited by E. N. Lee and M. Mandelbaum. London: Johns Hopkins Press. 1967.

Russell, Bertrand. *The Problems of Philosophy*. Oxford: Oxford University Press, 1959. (First published in 1912)

Russell, Bertrand. *Our Knowledge of the External World*. Chicago: The Open Court,1914.

Russell, Bertrand. "The Relation of Sense-Data to Physics." In *Mysticism and Logic and Other Essays*. London: George Allen & Unwin, 1963. (First published in 1917)

Russell, Bertrand. "Knowledge by Acquaintance and Knowledge by Description." In *Mysticism and Logic and Other Essays*. London: George Allen & Unwin, 1963. (First published in 1917)

Russell, Bertrand. *Introduction to Mathematical Philosophy*, 2nd ed. London: George Allen & Unwin, 1920.

Russell, Bertrand. "On the Nature of Acquaintance." In *Logic and Knowledge*. Edited by Robert C. Marsh. London: George Allen & Unwin, 1956.

Russell, Bertrand. "Philosophy of Logical Atomism." In *Logic and Knowledge*. Edited by Robert C. Marsh. London: George Allen & Unwin, 1956.

Russell, Bertrand. *Theory of Knowledge: The 1913 Manuscript—The Collected Papers of Bertrand Russell* Vol.7. Edited by Elizabeth Eames. London: George Allen & Unwin, 1984.

Schopenhauer, Arthur. *The World as Will and Representation*. 2 vols. Translated by E. F. J. Payne. New York: Dover, 1969.

Smith, David W. and Ronald McIntyre. *Husserl and Intentionality*. Dordrecht: D. Reidel, 1982.

Soulez, Antonia. "Wittgenstein and Phenomenology or: Two Languages for One Wittgenstein." In *Wittgenstein in Focus: Grazer Philosophische Studien* 33/34 (1989).

Spiegelberg, Herbert. "The Puzzle of Wittgenstein's *Phänomenologie* (1929-?) (with Supplement 1979)." In *The Context of the Phenomenological Movement*. Edited by Herbert Spiegelberg. The Hague: Martinus Nijhoff, 1981.

Spiegelberg, Herbert. *The Phenomenological Movement*. 2nd and 3rd eds. The Hague: Martinus Nijhoff, 1965 and 1982.

Stern, David. "Review Essay: Recent Work on Wittgenstein, 1980-1990." *Synthese* 98 (1994): 415-58.

Stern, David. *Wittgenstein on Mind and Language*. Oxford: Oxford University Press, 1995.

Stern, David. "Availability of Wittgenstein's Philosophy." In *The Cambridge Companion to Wittgenstein*. Edited by Hans Sluga and David G. Stern. Cambridge: Cambridge University Press. 1996.

Ströker, Elisabeth. *Husserl's Transcendental Phenomenology*. Translated by Lee Hardy. Stanford: Stanford University Press, 1993.

Thompson, Evan. "Novel Colours" *Philosophical Studies* 68 (1992): 321-50.

Tragesser, Robert S. *Phenomenology and Logic*. Ithaca: Cornell University Press, 1977.

Uebel, Thomas. "Physicalism in Wittgenstein and The Vienna Circle." In *Physics, Philosophy and the Scientific Community*. Edited by K. Gavroglu et al. Dordrecht: Kluwer Academic Publishers. 1995.

Urmson, J. O. *Philosophical Analysis*. Oxford: The Clarendon Press, 1956.

Urmson, J. O. "Russell on Acquaintance with the Past." *Philosophical Review* (1969): 510-15.

Vaina, Lucia M. "'What' and 'Where' in the Human Visual System: Two Hierarchies of Visual Modules." *Synthese* 83 (1990): 49-91.

van Heijenoort, Jean. "Logic as Calculus and Logic as Language." *Synthese* 17 (1967): 324-30.

van Peursen, C. A. "Edmund Husserl and Ludwig Wittgenstein." *Philosophy and Phenomenological Research* 20 (1959): 181-97.

Visser, Henk. "Wittgenstein's Debt to Mach's *Popular Scientific Lectures*." *Mind* 91 (1982): 102-5.

von Wright, G. H. *Wittgenstein*. Minneapolis: University of Minnesota Press, 1982.

Vossenkuhl, Wilhelm. "Farben in Carnaps *Logischen Aufbau*." In *Wissenschaft und Subjektivität*. Edited by David Bell and Wilhelm Vossenkuhl. Berlin: Akademie Verlag, 1992.

Westphal, Jonathan. *Colour: A Philosophical Introduction*. 2nd ed. Oxford: Basil Blackwell, 1991.

Wilson, Andrew. "Hertz, Boltzmann and Wittgenstein Reconsidered." *Studies in History and Philosophy of Science* 20 (1989): 245-63.

INDEX

22. P. Suppes, *Studies in the Methodology and Foundations of Science. Selected Papers from 1951 to 1969*. 1969 ISBN 90-277-0020-6
23. J. Hintikka, *Models for Modalities*. Selected Essays. 1969 ISBN 90-277-0078-8; Pb 90-277-0598-4
24. N. Rescher *et al.* (eds.), *Essays in Honor of Carl G. Hempel*. A Tribute on the Occasion of His 65th Birthday. 1969 ISBN 90-277-0085-0
25. P. V. Tavanec (ed.), *Problems of the Logic of Scientific Knowledge*. Translated from Russian. 1970 ISBN 90-277-0087-7
26. M. Swain (ed.), *Induction, Acceptance, and Rational Belief*. 1970 ISBN 90-277-0086-9
27. R. S. Cohen and R. J. Seeger (eds.), *Ernst Mach: Physicist and Philosopher*. [Boston Studies in the Philosophy of Science, Vol. VI]. 1970 ISBN 90-277-0016-8
28. J. Hintikka and P. Suppes, *Information and Inference*. 1970 ISBN 90-277-0155-5
29. K. Lambert, *Philosophical Problems in Logic*. Some Recent Developments. 1970 ISBN 90-277-0079-6
30. R. A. Eberle, *Nominalistic Systems*. 1970 ISBN 90-277-0161-X
31. P. Weingartner and G. Zecha (eds.), *Induction, Physics, and Ethics*. 1970 ISBN 90-277-0158-X
32. E. W. Beth, *Aspects of Modern Logic*. Translated from Dutch. 1970 ISBN 90-277-0173-3
33. R. Hilpinen (ed.), *Deontic Logic*. Introductory and Systematic Readings. 1971 *See also* No. 152. ISBN Pb (1981 rev.) 90-277-1302-2
34. J.-L. Krivine, *Introduction to Axiomatic Set Theory*. Translated from French. 1971 ISBN 90-277-0169-5; Pb 90-277-0411-2
35. J. D. Sneed, *The Logical Structure of Mathematical Physics*. 2nd rev. ed., 1979 ISBN 90-277-1056-2; Pb 90-277-1059-7
36. C. R. Kordig, *The Justification of Scientific Change*. 1971 ISBN 90-277-0181-4; Pb 90-277-0475-9
37. M. Čapek, *Bergson and Modern Physics*. A Reinterpretation and Re-evaluation. [Boston Studies in the Philosophy of Science, Vol. VII] 1971 ISBN 90-277-0186-5
38. N. R. Hanson, *What I Do Not Believe, and Other Essays*. Ed. by S. Toulmin and H. Woolf. 1971 ISBN 90-277-0191-1
39. R. C. Buck and R. S. Cohen (eds.), *PSA 1970*. Proceedings of the Second Biennial Meeting of the Philosophy of Science Association, Boston, Fall 1970. In Memory of Rudolf Carnap. [Boston Studies in the Philosophy of Science, Vol. VIII] 1971 ISBN 90-277-0187-3; Pb 90-277-0309-4
40. D. Davidson and G. Harman (eds.), *Semantics of Natural Language*. 1972 ISBN 90-277-0304-3; Pb 90-277-0310-8
41. Y. Bar-Hillel (ed.), *Pragmatics of Natural Languages*. 1971 ISBN 90-277-0194-6; Pb 90-277-0599-2
42. S. Stenlund, *Combinators, λ-Terms and Proof Theory*. 1972 ISBN 90-277-0305-1
43. M. Strauss, *Modern Physics and Its Philosophy*. Selected Paper in the Logic, History, and Philosophy of Science. 1972 ISBN 90-277-0230-6
44. M. Bunge, *Method, Model and Matter*. 1973 ISBN 90-277-0252-7
45. M. Bunge, *Philosophy of Physics*. 1973 ISBN 90-277-0253-5
46. A. A. Zinov'ev, *Foundations of the Logical Theory of Scientific Knowledge (Complex Logic)*. Revised and enlarged English edition with an appendix by G. A. Smirnov, E. A. Sidorenka, A. M. Fedina and L. A. Bobrova. [Boston Studies in the Philosophy of Science, Vol. IX] 1973 ISBN 90-277-0193-8; Pb 90-277-0324-8
47. L. Tondl, *Scientific Procedures*. A Contribution concerning the Methodological Problems of Scientific Concepts and Scientific Explanation. Translated from Czech by D. Short. Edited by R.S. Cohen and M.W. Wartofsky. [Boston Studies in the Philosophy of Science, Vol. X] 1973 ISBN 90-277-0147-4; Pb 90-277-0323-X

48. N. R. Hanson, *Constellations and Conjectures*. 1973 ISBN 90-277-0192-X
49. K. J. J. Hintikka, J. M. E. Moravcsik and P. Suppes (eds.), *Approaches to Natural Language*. 1973 ISBN 90-277-0220-9; Pb 90-277-0233-0
50. M. Bunge (ed.), *Exact Philosophy*. Problems, Tools and Goals. 1973 ISBN 90-277-0251-9
51. R. J. Bogdan and I. Niiniluoto (eds.), *Logic, Language and Probability*. 1973 ISBN 90-277-0312-4
52. G. Pearce and P. Maynard (eds.), *Conceptual Change*. 1973 ISBN 90-277-0287-X; Pb 90-277-0339-6
53. I. Niiniluoto and R. Tuomela, *Theoretical Concepts and Hypothetico-inductive Inference*. 1973 ISBN 90-277-0343-4
54. R. Fraissé, *Course of Mathematical Logic* – Volume 1: *Relation and Logical Formula*. Translated from French. 1973 ISBN 90-277-0268-3; Pb 90-277-0403-1 (For *Volume 2* see under No. 69).
55. A. Grünbaum, *Philosophical Problems of Space and Time*. Edited by R.S. Cohen and M.W. Wartofsky. 2nd enlarged ed. [Boston Studies in the Philosophy of Science, Vol. XII] 1973 ISBN 90-277-0357-4; Pb 90-277-0358-2
56. P. Suppes (ed.), *Space, Time and Geometry*. 1973 ISBN 90-277-0386-8; Pb 90-277-0442-2
57. H. Kelsen, *Essays in Legal and Moral Philosophy*. Selected and introduced by O. Weinberger. Translated from German by P. Heath. 1973 ISBN 90-277-0388-4
58. R. J. Seeger and R. S. Cohen (eds.), *Philosophical Foundations of Science*. [Boston Studies in the Philosophy of Science, Vol. XI] 1974 ISBN 90-277-0390-6; Pb 90-277-0376-0
59. R. S. Cohen and M. W. Wartofsky (eds.), *Logical and Epistemological Studies in Contemporary Physics*. [Boston Studies in the Philosophy of Science, Vol. XIII] 1973 ISBN 90-277-0391-4; Pb 90-277-0377-9
60. R. S. Cohen and M. W. Wartofsky (eds.), *Methodological and Historical Essays in the Natural and Social Sciences. Proceedings of the Boston Colloquium for the Philosophy of Science, 1969–1972*. [Boston Studies in the Philosophy of Science, Vol. XIV] 1974 ISBN 90-277-0392-2; Pb 90-277-0378-7
61. R. S. Cohen, J. J. Stachel and M. W. Wartofsky (eds.), *For Dirk Struik. Scientific, Historical and Political Essays*. [Boston Studies in the Philosophy of Science, Vol. XV] 1974 ISBN 90-277-0393-0; Pb 90-277-0379-5
62. K. Ajdukiewicz, *Pragmatic Logic*. Translated from Polish by O. Wojtasiewicz. 1974 ISBN 90-277-0326-4
63. S. Stenlund (ed.), *Logical Theory and Semantic Analysis*. Essays dedicated to Stig Kanger on His 50th Birthday. 1974 ISBN 90-277-0438-4
64. K. F. Schaffner and R. S. Cohen (eds.), *PSA 1972. Proceedings of the Third Biennial Meeting of the Philosophy of Science Association*. [Boston Studies in the Philosophy of Science, Vol. XX] 1974 ISBN 90-277-0408-2; Pb 90-277-0409-0
65. H. E. Kyburg, Jr., *The Logical Foundations of Statistical Inference*. 1974 ISBN 90-277-0330-2; Pb 90-277-0430-9
66. M. Grene, *The Understanding of Nature*. Essays in the Philosophy of Biology. [Boston Studies in the Philosophy of Science, Vol. XXIII] 1974 ISBN 90-277-0462-7; Pb 90-277-0463-5
67. J. M. Broekman, *Structuralism: Moscow, Prague, Paris*. Translated from German. 1974 ISBN 90-277-0478-3
68. N. Geschwind, *Selected Papers on Language and the Brain*. [Boston Studies in the Philosophy of Science, Vol. XVI] 1974 ISBN 90-277-0262-4; Pb 90-277-0263-2
69. R. Fraissé, *Course of Mathematical Logic* – Volume 2: *Model Theory*. Translated from French. 1974 ISBN 90-277-0269-1; Pb 90-277-0510-0 (For *Volume 1* see under No. 54)

70. A. Grzegorczyk, *An Outline of Mathematical Logic*. Fundamental Results and Notions explained with all Details. Translated from Polish. 1974
ISBN 90-277-0359-0; Pb 90-277-0447-3
71. F. von Kutschera, *Philosophy of Language*. 1975 ISBN 90-277-0591-7
72. J. Manninen and R. Tuomela (eds.), *Essays on Explanation and Understanding*. Studies in the Foundations of Humanities and Social Sciences. 1976 ISBN 90-277-0592-5
73. J. Hintikka (ed.), *Rudolf Carnap, Logical Empiricist*. Materials and Perspectives. 1975
ISBN 90-277-0583-6
74. M. Čapek (ed.), *The Concepts of Space and Time*. Their Structure and Their Development. [Boston Studies in the Philosophy of Science, Vol. XXII] 1976
ISBN 90-277-0355-8; Pb 90-277-0375-2
75. J. Hintikka and U. Remes, *The Method of Analysis*. Its Geometrical Origin and Its General Significance. [Boston Studies in the Philosophy of Science, Vol. XXV] 1974
ISBN 90-277-0532-1; Pb 90-277-0543-7
76. J. E. Murdoch and E. D. Sylla (eds.), *The Cultural Context of Medieval Learning*. [Boston Studies in the Philosophy of Science, Vol. XXVI] 1975
ISBN 90-277-0560-7; Pb 90-277-0587-9
77. S. Amsterdamski, *Between Experience and Metaphysics*. Philosophical Problems of the Evolution of Science. [Boston Studies in the Philosophy of Science, Vol. XXXV] 1975
ISBN 90-277-0568-2; Pb 90-277-0580-1
78. P. Suppes (ed.), *Logic and Probability in Quantum Mechanics*. 1976
ISBN 90-277-0570-4; Pb 90-277-1200-X
79. H. von Helmholtz: *Epistemological Writings. The Paul Hertz / Moritz Schlick Centenary Edition of 1921 with Notes and Commentary by the Editors*. Newly translated from German by M. F. Lowe. Edited, with an Introduction and Bibliography, by R. S. Cohen and Y. Elkana. [Boston Studies in the Philosophy of Science, Vol. XXXVII] 1975
ISBN 90-277-0290-X; Pb 90-277-0582-8
80. J. Agassi, *Science in Flux*. [Boston Studies in the Philosophy of Science, Vol. XXVIII] 1975 ISBN 90-277-0584-4; Pb 90-277-0612-2
81. S. G. Harding (ed.), *Can Theories Be Refuted?* Essays on the Duhem-Quine Thesis. 1976
ISBN 90-277-0629-8; Pb 90-277-0630-1
82. S. Nowak, *Methodology of Sociological Research*. General Problems. 1977
ISBN 90-277-0486-4
83. J. Piaget, J.-B. Grize, A. Szemińska and V. Bang, *Epistemology and Psychology of Functions*. Translated from French. 1977 ISBN 90-277-0804-5
84. M. Grene and E. Mendelsohn (eds.), *Topics in the Philosophy of Biology*. [Boston Studies in the Philosophy of Science, Vol. XXVII] 1976
ISBN 90-277-0595-X; Pb 90-277-0596-8
85. E. Fischbein, *The Intuitive Sources of Probabilistic Thinking in Children*. 1975
ISBN 90-277-0626-3; Pb 90-277-1190-9
86. E. W. Adams, *The Logic of Conditionals*. An Application of Probability to Deductive Logic. 1975 ISBN 90-277-0631-X
87. M. Przełęcki and R. Wójcicki (eds.), *Twenty-Five Years of Logical Methodology in Poland*. Translated from Polish. 1976 ISBN 90-277-0601-8
88. J. Topolski, *The Methodology of History*. Translated from Polish by O. Wojtasiewicz. 1976 ISBN 90-277-0550-X
89. A. Kasher (ed.), *Language in Focus: Foundations, Methods and Systems*. Essays dedicated to Yehoshua Bar-Hillel. [Boston Studies in the Philosophy of Science, Vol. XLIII] 1976 ISBN 90-277-0644-1; Pb 90-277-0645-X
90. J. Hintikka, *The Intentions of Intentionality and Other New Models for Modalities*. 1975
ISBN 90-277-0633-6; Pb 90-277-0634-4

91. W. Stegmüller, *Collected Papers on Epistemology, Philosophy of Science and History of Philosophy*. 2 Volumes. 1977 Set ISBN 90-277-0767-7
92. D. M. Gabbay, *Investigations in Modal and Tense Logics with Applications to Problems in Philosophy and Linguistics*. 1976 ISBN 90-277-0656-5
93. R. J. Bogdan, *Local Induction*. 1976 ISBN 90-277-0649-2
94. S. Nowak, *Understanding and Prediction*. Essays in the Methodology of Social and Behavioral Theories. 1976 ISBN 90-277-0558-5; Pb 90-277-1199-2
95. P. Mittelstaedt, *Philosophical Problems of Modern Physics*. [Boston Studies in the Philosophy of Science, Vol. XVIII] 1976 ISBN 90-277-0285-3; Pb 90-277-0506-2
96. G. Holton and W. A. Blanpied (eds.), *Science and Its Public: The Changing Relationship*. [Boston Studies in the Philosophy of Science, Vol. XXXIII] 1976 ISBN 90-277-0657-3; Pb 90-277-0658-1
97. M. Brand and D. Walton (eds.), *Action Theory*. 1976 ISBN 90-277-0671-9
98. P. Gochet, *Outline of a Nominalist Theory of Propositions*. An Essay in the Theory of Meaning and in the Philosophy of Logic. 1980 ISBN 90-277-1031-7
99. R. S. Cohen, P. K. Feyerabend, and M. W. Wartofsky (eds.), *Essays in Memory of Imre Lakatos*. [Boston Studies in the Philosophy of Science, Vol. XXXIX] 1976 ISBN 90-277-0654-9; Pb 90-277-0655-7
100. R. S. Cohen and J. J. Stachel (eds.), *Selected Papers of Léon Rosenfield*. [Boston Studies in the Philosophy of Science, Vol. XXI] 1979 ISBN 90-277-0651-4; Pb 90-277-0652-2
101. R. S. Cohen, C. A. Hooker, A. C. Michalos and J. W. van Evra (eds.), *PSA 1974. Proceedings of the 1974 Biennial Meeting of the Philosophy of Science Association*. [Boston Studies in the Philosophy of Science, Vol. XXXII] 1976 ISBN 90-277-0647-6; Pb 90-277-0648-4
102. Y. Fried and J. Agassi, *Paranoia*. A Study in Diagnosis. [Boston Studies in the Philosophy of Science, Vol. L] 1976 ISBN 90-277-0704-9; Pb 90-277-0705-7
103. M. Przełęcki, K. Szaniawski and R. Wójcicki (eds.), *Formal Methods in the Methodology of Empirical Sciences*. 1976 ISBN 90-277-0698-0
104. J. M. Vickers, *Belief and Probability*. 1976 ISBN 90-277-0744-8
105. K. H. Wolff, *Surrender and Catch*. Experience and Inquiry Today. [Boston Studies in the Philosophy of Science, Vol. LI] 1976 ISBN 90-277-0758-8; Pb 90-277-0765-0
106. K. Kosík, *Dialectics of the Concrete*. A Study on Problems of Man and World. [Boston Studies in the Philosophy of Science, Vol. LII] 1976 ISBN 90-277-0761-8; Pb 90-277-0764-2
107. N. Goodman, *The Structure of Appearance*. 3rd ed. with an Introduction by G. Hellman. [Boston Studies in the Philosophy of Science, Vol. LIII] 1977 ISBN 90-277-0773-1; Pb 90-277-0774-X
108. K. Ajdukiewicz, *The Scientific World-Perspective and Other Essays, 1931-1963*. Translated from Polish. Edited and with an Introduction by J. Giedymin. 1978 ISBN 90-277-0527-5
109. R. L. Causey, *Unity of Science*. 1977 ISBN 90-277-0779-0
110. R. E. Grandy, *Advanced Logic for Applications*. 1977 ISBN 90-277-0781-2
111. R. P. McArthur, *Tense Logic*. 1976 ISBN 90-277-0697-2
112. L. Lindahl, *Position and Change*. A Study in Law and Logic. Translated from Swedish by P. Needham. 1977 ISBN 90-277-0787-1
113. R. Tuomela, *Dispositions*. 1978 ISBN 90-277-0810-X
114. H. A. Simon, *Models of Discovery and Other Topics in the Methods of Science*. [Boston Studies in the Philosophy of Science, Vol. LIV] 1977 ISBN 90-277-0812-6; Pb 90-277-0858-4
115. R. D. Rosenkrantz, *Inference, Method and Decision*. Towards a Bayesian Philosophy of Science. 1977 ISBN 90-277-0817-7; Pb 90-277-0818-5

116. R. Tuomela, *Human Action and Its Explanation.* A Study on the Philosophical Foundations of Psychology. 1977 ISBN 90-277-0824-X
117. M. Lazerowitz, *The Language of Philosophy.* Freud and Wittgenstein. [Boston Studies in the Philosophy of Science, Vol. LV] 1977 ISBN 90-277-0826-6; Pb 90-277-0862-2
118. Not published
119. J. Pelc (ed.), *Semiotics in Poland, 1894–1969.* Translated from Polish. 1979 ISBN 90-277-0811-8
120. I. Pörn, *Action Theory and Social Science.* Some Formal Models. 1977 ISBN 90-277-0846-0
121. J. Margolis, *Persons and Mind.* The Prospects of Nonreductive Materialism. [Boston Studies in the Philosophy of Science, Vol. LVII] 1977 ISBN 90-277-0854-1; Pb 90-277-0863-0
122. J. Hintikka, I. Niiniluoto, and E. Saarinen (eds.), *Essays on Mathematical and Philosophical Logic.* 1979 ISBN 90-277-0879-7
123. T. A. F. Kuipers, *Studies in Inductive Probability and Rational Expectation.* 1978 ISBN 90-277-0882-7
124. E. Saarinen, R. Hilpinen, I. Niiniluoto and M. P. Hintikka (eds.), *Essays in Honour of Jaakko Hintikka on the Occasion of His 50th Birthday.* 1979 ISBN 90-277-0916-5
125. G. Radnitzky and G. Andersson (eds.), *Progress and Rationality in Science.* [Boston Studies in the Philosophy of Science, Vol. LVIII] 1978 ISBN 90-277-0921-1; Pb 90-277-0922-X
126. P. Mittelstaedt, *Quantum Logic.* 1978 ISBN 90-277-0925-4
127. K. A. Bowen, *Model Theory for Modal Logic.* Kripke Models for Modal Predicate Calculi. 1979 ISBN 90-277-0929-7
128. H. A. Bursen, *Dismantling the Memory Machine.* A Philosophical Investigation of Machine Theories of Memory. 1978 ISBN 90-277-0933-5
129. M. W. Wartofsky, *Models.* Representation and the Scientific Understanding. [Boston Studies in the Philosophy of Science, Vol. XLVIII] 1979 ISBN 90-277-0736-7; Pb 90-277-0947-5
130. D. Ihde, *Technics and Praxis.* A Philosophy of Technology. [Boston Studies in the Philosophy of Science, Vol. XXIV] 1979 ISBN 90-277-0953-X; Pb 90-277-0954-8
131. J. J. Wiatr (ed.), *Polish Essays in the Methodology of the Social Sciences.* [Boston Studies in the Philosophy of Science, Vol. XXIX] 1979 ISBN 90-277-0723-5; Pb 90-277-0956-4
132. W. C. Salmon (ed.), *Hans Reichenbach: Logical Empiricist.* 1979 ISBN 90-277-0958-0
133. P. Bieri, R.-P. Horstmann and L. Krüger (eds.), *Transcendental Arguments in Science.* Essays in Epistemology. 1979 ISBN 90-277-0963-7; Pb 90-277-0964-5
134. M. Marković and G. Petrović (eds.), *Praxis.* Yugoslav Essays in the Philosophy and Methodology of the Social Sciences. [Boston Studies in the Philosophy of Science, Vol. XXXVI] 1979 ISBN 90-277-0727-8; Pb 90-277-0968-8
135. R. Wójcicki, *Topics in the Formal Methodology of Empirical Sciences.* Translated from Polish. 1979 ISBN 90-277-1004-X
136. G. Radnitzky and G. Andersson (eds.), *The Structure and Development of Science.* [Boston Studies in the Philosophy of Science, Vol. LIX] 1979 ISBN 90-277-0994-7; Pb 90-277-0995-5
137. J. C. Webb, *Mechanism, Mentalism and Metamathematics.* An Essay on Finitism. 1980 ISBN 90-277-1046-5
138. D. F. Gustafson and B. L. Tapscott (eds.), *Body, Mind and Method.* Essays in Honor of Virgil C. Aldrich. 1979 ISBN 90-277-1013-9
139. L. Nowak, *The Structure of Idealization.* Towards a Systematic Interpretation of the Marxian Idea of Science. 1980 ISBN 90-277-1014-7

140. C. Perelman, *The New Rhetoric and the Humanities.* Essays on Rhetoric and Its Applications. Translated from French and German. With an Introduction by H. Zyskind. 1979 ISBN 90-277-1018-X; Pb 90-277-1019-8
141. W. Rabinowicz, *Universalizability.* A Study in Morals and Metaphysics. 1979 ISBN 90-277-1020-2
142. C. Perelman, *Justice, Law and Argument.* Essays on Moral and Legal Reasoning. Translated from French and German. With an Introduction by H.J. Berman. 1980 ISBN 90-277-1089-9; Pb 90-277-1090-2
143. S. Kanger and S. Öhman (eds.), *Philosophy and Grammar.* Papers on the Occasion of the Quincentennial of Uppsala University. 1981 ISBN 90-277-1091-0
144. T. Pawlowski, *Concept Formation in the Humanities and the Social Sciences.* 1980 ISBN 90-277-1096-1
145. J. Hintikka, D. Gruender and E. Agazzi (eds.), *Theory Change, Ancient Axiomatics and Galileo's Methodology.* Proceedings of the 1978 Pisa Conference on the History and Philosophy of Science, Volume I. 1981 ISBN 90-277-1126-7
146. J. Hintikka, D. Gruender and E. Agazzi (eds.), *Probabilistic Thinking, Thermodynamics, and the Interaction of the History and Philosophy of Science.* Proceedings of the 1978 Pisa Conference on the History and Philosophy of Science, Volume II. 1981 ISBN 90-277-1127-5
147. U. Mönnich (ed.), *Aspects of Philosophical Logic.* Some Logical Forays into Central Notions of Linguistics and Philosophy. 1981 ISBN 90-277-1201-8
148. D. M. Gabbay, *Semantical Investigations in Heyting's Intuitionistic Logic.* 1981 ISBN 90-277-1202-6
149. E. Agazzi (ed.), *Modern Logic – A Survey.* Historical, Philosophical, and Mathematical Aspects of Modern Logic and Its Applications. 1981 ISBN 90-277-1137-2
150. A. F. Parker-Rhodes, *The Theory of Indistinguishables.* A Search for Explanatory Principles below the Level of Physics. 1981 ISBN 90-277-1214-X
151. J. C. Pitt, *Pictures, Images, and Conceptual Change.* An Analysis of Wilfrid Sellars' Philosophy of Science. 1981 ISBN 90-277-1276-X; Pb 90-277-1277-8
152. R. Hilpinen (ed.), *New Studies in Deontic Logic.* Norms, Actions, and the Foundations of Ethics. 1981 ISBN 90-277-1278-6; Pb 90-277-1346-4
153. C. Dilworth, *Scientific Progress.* A Study Concerning the Nature of the Relation between Successive Scientific Theories. 3rd rev. ed., 1994 ISBN 0-7923-2487-0; Pb 0-7923-2488-9
154. D. Woodruff Smith and R. McIntyre, *Husserl and Intentionality.* A Study of Mind, Meaning, and Language. 1982 ISBN 90-277-1392-8; Pb 90-277-1730-3
155. R. J. Nelson, *The Logic of Mind.* 2nd. ed., 1989 ISBN 90-277-2819-4; Pb 90-277-2822-4
156. J. F. A. K. van Benthem, *The Logic of Time.* A Model-Theoretic Investigation into the Varieties of Temporal Ontology, and Temporal Discourse. 1983; 2nd ed., 1991 ISBN 0-7923-1081-0
157. R. Swinburne (ed.), *Space, Time and Causality.* 1983 ISBN 90-277-1437-1
158. E. T. Jaynes, *Papers on Probability, Statistics and Statistical Physics.* Ed. by R. D. Rozenkrantz. 1983 ISBN 90-277-1448-7; Pb (1989) 0-7923-0213-3
159. T. Chapman, *Time: A Philosophical Analysis.* 1982 ISBN 90-277-1465-7
160. E. N. Zalta, *Abstract Objects.* An Introduction to Axiomatic Metaphysics. 1983 ISBN 90-277-1474-6
161. S. Harding and M. B. Hintikka (eds.), *Discovering Reality.* Feminist Perspectives on Epistemology, Metaphysics, Methodology, and Philosophy of Science. 1983 ISBN 90-277-1496-7; Pb 90-277-1538-6
162. M. A. Stewart (ed.), *Law, Morality and Rights.* 1983 ISBN 90-277-1519-X

163. D. Mayr and G. Süssmann (eds.), *Space, Time, and Mechanics*. Basic Structures of a Physical Theory. 1983 ISBN 90-277-1525-4
164. D. Gabbay and F. Guenthner (eds.), *Handbook of Philosophical Logic*. Vol. I: Elements of Classical Logic. 1983 ISBN 90-277-1542-4
165. D. Gabbay and F. Guenthner (eds.), *Handbook of Philosophical Logic*. Vol. II: Extensions of Classical Logic. 1984 ISBN 90-277-1604-8
166. D. Gabbay and F. Guenthner (eds.), *Handbook of Philosophical Logic*. Vol. III: Alternative to Classical Logic. 1986 ISBN 90-277-1605-6
167. D. Gabbay and F. Guenthner (eds.), *Handbook of Philosophical Logic*. Vol. IV: Topics in the Philosophy of Language. 1989 ISBN 90-277-1606-4
168. A. J. I. Jones, *Communication and Meaning*. An Essay in Applied Modal Logic. 1983 ISBN 90-277-1543-2
169. M. Fitting, *Proof Methods for Modal and Intuitionistic Logics*. 1983 ISBN 90-277-1573-4
170. J. Margolis, *Culture and Cultural Entities*. Toward a New Unity of Science. 1984 ISBN 90-277-1574-2
171. R. Tuomela, *A Theory of Social Action*. 1984 ISBN 90-277-1703-6
172. J. J. E. Gracia, E. Rabossi, E. Villanueva and M. Dascal (eds.), *Philosophical Analysis in Latin America*. 1984 ISBN 90-277-1749-4
173. P. Ziff, *Epistemic Analysis*. A Coherence Theory of Knowledge. 1984 ISBN 90-277-1751-7
174. P. Ziff, *Antiaesthetics*. An Appreciation of the Cow with the Subtile Nose. 1984 ISBN 90-277-1773-7
175. W. Balzer, D. A. Pearce, and H.-J. Schmidt (eds.), *Reduction in Science*. Structure, Examples, Philosophical Problems. 1984 ISBN 90-277-1811-3
176. A. Peczenik, L. Lindahl and B. van Roermund (eds.), *Theory of Legal Science*. Proceedings of the Conference on Legal Theory and Philosophy of Science (Lund, Sweden, December 1983). 1984 ISBN 90-277-1834-2
177. I. Niiniluoto, *Is Science Progressive?* 1984 ISBN 90-277-1835-0
178. B. K. Matilal and J. L. Shaw (eds.), *Analytical Philosophy in Comparative Perspective*. Exploratory Essays in Current Theories and Classical Indian Theories of Meaning and Reference. 1985 ISBN 90-277-1870-9
179. P. Kroes, *Time: Its Structure and Role in Physical Theories*. 1985 ISBN 90-277-1894-6
180. J. H. Fetzer, *Sociobiology and Epistemology*. 1985 ISBN 90-277-2005-3; Pb 90-277-2006-1
181. L. Haaparanta and J. Hintikka (eds.), *Frege Synthesized*. Essays on the Philosophical and Foundational Work of Gottlob Frege. 1986 ISBN 90-277-2126-2
182. M. Detlefsen, *Hilbert's Program*. An Essay on Mathematical Instrumentalism. 1986 ISBN 90-277-2151-3
183. J. L. Golden and J. J. Pilotta (eds.), *Practical Reasoning in Human Affairs*. Studies in Honor of Chaim Perelman. 1986 ISBN 90-277-2255-2
184. H. Zandvoort, *Models of Scientific Development and the Case of Nuclear Magnetic Resonance*. 1986 ISBN 90-277-2351-6
185. I. Niiniluoto, *Truthlikeness*. 1987 ISBN 90-277-2354-0
186. W. Balzer, C. U. Moulines and J. D. Sneed, *An Architectonic for Science*. The Structuralist Program. 1987 ISBN 90-277-2403-2
187. D. Pearce, *Roads to Commensurability*. 1987 ISBN 90-277-2414-8
188. L. M. Vaina (ed.), *Matters of Intelligence*. Conceptual Structures in Cognitive Neuroscience. 1987 ISBN 90-277-2460-1
189. H. Siegel, *Relativism Refuted*. A Critique of Contemporary Epistemological Relativism. 1987 ISBN 90-277-2469-5

190. W. Callebaut and R. Pinxten, *Evolutionary Epistemology*. A Multiparadigm Program, with a Complete Evolutionary Epistemology Bibliograph. 1987 ISBN 90-277-2582-9
191. J. Kmita, *Problems in Historical Epistemology*. 1988 ISBN 90-277-2199-8
192. J. H. Fetzer (ed.), *Probability and Causality*. Essays in Honor of Wesley C. Salmon, with an Annotated Bibliography. 1988 ISBN 90-277-2607-8; Pb 1-5560-8052-2
193. A. Donovan, L. Laudan and R. Laudan (eds.), *Scrutinizing Science*. Empirical Studies of Scientific Change. 1988 ISBN 90-277-2608-6
194. H.R. Otto and J.A. Tuedio (eds.), *Perspectives on Mind*. 1988 ISBN 90-277-2640-X
195. D. Batens and J.P. van Bendegem (eds.), *Theory and Experiment*. Recent Insights and New Perspectives on Their Relation. 1988 ISBN 90-277-2645-0
196. J. Österberg, *Self and Others*. A Study of Ethical Egoism. 1988 ISBN 90-277-2648-5
197. D.H. Helman (ed.), *Analogical Reasoning*. Perspectives of Artificial Intelligence, Cognitive Science, and Philosophy. 1988 ISBN 90-277-2711-2
198. J. Wolenski, *Logic and Philosophy in the Lvov-Warsaw School*. 1989 ISBN 90-277-2749-X
199. R. Wójcicki, *Theory of Logical Calculi*. Basic Theory of Consequence Operations. 1988 ISBN 90-277-2785-6
200. J. Hintikka and M.B. Hintikka, *The Logic of Epistemology and the Epistemology of Logic*. Selected Essays. 1989 ISBN 0-7923-0040-8; Pb 0-7923-0041-6
201. E. Agazzi (ed.), *Probability in the Sciences*. 1988 ISBN 90-277-2808-9
202. M. Meyer (ed.), *From Metaphysics to Rhetoric*. 1989 ISBN 90-277-2814-3
203. R.L. Tieszen, *Mathematical Intuition*. Phenomenology and Mathematical Knowledge. 1989 ISBN 0-7923-0131-5
204. A. Melnick, *Space, Time, and Thought in Kant*. 1989 ISBN 0-7923-0135-8
205. D.W. Smith, *The Circle of Acquaintance*. Perception, Consciousness, and Empathy. 1989 ISBN 0-7923-0252-4
206. M.H. Salmon (ed.), *The Philosophy of Logical Mechanism*. Essays in Honor of Arthur W. Burks. With his Responses, and with a Bibliography of Burk's Work. 1990 ISBN 0-7923-0325-3
207. M. Kusch, *Language as Calculus vs. Language as Universal Medium*. A Study in Husserl, Heidegger, and Gadamer. 1989 ISBN 0-7923-0333-4
208. T.C. Meyering, *Historical Roots of Cognitive Science*. The Rise of a Cognitive Theory of Perception from Antiquity to the Nineteenth Century. 1989 ISBN 0-7923-0349-0
209. P. Kosso, *Observability and Observation in Physical Science*. 1989 ISBN 0-7923-0389-X
210. J. Kmita, *Essays on the Theory of Scientific Cognition*. 1990 ISBN 0-7923-0441-1
211. W. Sieg (ed.), *Acting and Reflecting*. The Interdisciplinary Turn in Philosophy. 1990 ISBN 0-7923-0512-4
212. J. Karpiński, *Causality in Sociological Research*. 1990 ISBN 0-7923-0546-9
213. H.A. Lewis (ed.), *Peter Geach: Philosophical Encounters*. 1991 ISBN 0-7923-0823-9
214. M. Ter Hark, *Beyond the Inner and the Outer*. Wittgenstein's Philosophy of Psychology. 1990 ISBN 0-7923-0850-6
215. M. Gosselin, *Nominalism and Contemporary Nominalism*. Ontological and Epistemological Implications of the Work of W.V.O. Quine and of N. Goodman. 1990 ISBN 0-7923-0904-9
216. J.H. Fetzer, D. Shatz and G. Schlesinger (eds.), *Definitions and Definability*. Philosophical Perspectives. 1991 ISBN 0-7923-1046-2
217. E. Agazzi and A. Cordero (eds.), *Philosophy and the Origin and Evolution of the Universe*. 1991 ISBN 0-7923-1322-4
218. M. Kusch, *Foucault's Strata and Fields*. An Investigation into Archaeological and Genealogical Science Studies. 1991 ISBN 0-7923-1462-X

219. C.J. Posy, *Kant's Philosophy of Mathematics.* Modern Essays. 1992 ISBN 0-7923-1495-6
220. G. Van de Vijver, *New Perspectives on Cybernetics.* Self-Organization, Autonomy and Connectionism. 1992 ISBN 0-7923-1519-7
221. J.C. Nyíri, *Tradition and Individuality.* Essays. 1992 ISBN 0-7923-1566-9
222. R. Howell, *Kant's Transcendental Deduction.* An Analysis of Main Themes in His Critical Philosophy. 1992 ISBN 0-7923-1571-5
223. A. García de la Sienra, *The Logical Foundations of the Marxian Theory of Value.* 1992 ISBN 0-7923-1778-5
224. D.S. Shwayder, *Statement and Referent.* An Inquiry into the Foundations of Our Conceptual Order. 1992 ISBN 0-7923-1803-X
225. M. Rosen, *Problems of the Hegelian Dialectic.* Dialectic Reconstructed as a Logic of Human Reality. 1993 ISBN 0-7923-2047-6
226. P. Suppes, *Models and Methods in the Philosophy of Science: Selected Essays.* 1993 ISBN 0-7923-2211-8
227. R. M. Dancy (ed.), *Kant and Critique: New Essays in Honor of W. H. Werkmeister.* 1993 ISBN 0-7923-2244-4
228. J. Woleński (ed.), *Philosophical Logic in Poland.* 1993 ISBN 0-7923-2293-2
229. M. De Rijke (ed.), *Diamonds and Defaults.* Studies in Pure and Applied Intensional Logic. 1993 ISBN 0-7923-2342-4
230. B.K. Matilal and A. Chakrabarti (eds.), *Knowing from Words.* Western and Indian Philosophical Analysis of Understanding and Testimony. 1994 ISBN 0-7923-2345-9
231. S.A. Kleiner, *The Logic of Discovery.* A Theory of the Rationality of Scientific Research. 1993 ISBN 0-7923-2371-8
232. R. Festa, *Optimum Inductive Methods.* A Study in Inductive Probability, Bayesian Statistics, and Verisimilitude. 1993 ISBN 0-7923-2460-9
233. P. Humphreys (ed.), *Patrick Suppes: Scientific Philosopher.* Vol. 1: Probability and Probabilistic Causality. 1994 ISBN 0-7923-2552-4
234. P. Humphreys (ed.), *Patrick Suppes: Scientific Philosopher.* Vol. 2: Philosophy of Physics, Theory Structure, and Measurement Theory. 1994 ISBN 0-7923-2553-2
235. P. Humphreys (ed.), *Patrick Suppes: Scientific Philosopher.* Vol. 3: Language, Logic, and Psychology. 1994 ISBN 0-7923-2862-0
Set ISBN (Vols 233–235) 0-7923-2554-0
236. D. Prawitz and D. Westerståhl (eds.), *Logic and Philosophy of Science in Uppsala.* Papers from the 9th International Congress of Logic, Methodology, and Philosophy of Science. 1994 ISBN 0-7923-2702-0
237. L. Haaparanta (ed.), *Mind, Meaning and Mathematics.* Essays on the Philosophical Views of Husserl and Frege. 1994 ISBN 0-7923-2703-9
238. J. Hintikka (ed.), *Aspects of Metaphor.* 1994 ISBN 0-7923-2786-1
239. B. McGuinness and G. Oliveri (eds.), *The Philosophy of Michael Dummett.* With Replies from Michael Dummett. 1994 ISBN 0-7923-2804-3
240. D. Jamieson (ed.), *Language, Mind, and Art.* Essays in Appreciation and Analysis, In Honor of Paul Ziff. 1994 ISBN 0-7923-2810-8
241. G. Preyer, F. Siebelt and A. Ulfig (eds.), *Language, Mind and Epistemology.* On Donald Davidson's Philosophy. 1994 ISBN 0-7923-2811-6
242. P. Ehrlich (ed.), *Real Numbers, Generalizations of the Reals, and Theories of Continua.* 1994 ISBN 0-7923-2689-X
243. G. Debrock and M. Hulswit (eds.), *Living Doubt.* Essays concerning the epistemology of Charles Sanders Peirce. 1994 ISBN 0-7923-2898-1
244. J. Srzednicki, *To Know or Not to Know.* Beyond Realism and Anti-Realism. 1994 ISBN 0-7923-2909-0
245. R. Egidi (ed.), *Wittgenstein: Mind and Language.* 1995 ISBN 0-7923-3171-0

246. A. Hyslop, *Other Minds*. 1995 ISBN 0-7923-3245-8
247. L. Pólos and M. Masuch (eds.), *Applied Logic: How, What and Why*. Logical Approaches to Natural Language. 1995 ISBN 0-7923-3432-9
248. M. Krynicki, M. Mostowski and L.M. Szczerba (eds.), *Quantifiers: Logics, Models and Computation*. Volume One: Surveys. 1995 ISBN 0-7923-3448-5
249. M. Krynicki, M. Mostowski and L.M. Szczerba (eds.), *Quantifiers: Logics, Models and Computation*. Volume Two: Contributions. 1995 ISBN 0-7923-3449-3
Set ISBN (Vols 248 + 249) 0-7923-3450-7
250. R.A. Watson, *Representational Ideas from Plato to Patricia Churchland*. 1995 ISBN 0-7923-3453-1
251. J. Hintikka (ed.), *From Dedekind to Gödel*. Essays on the Development of the Foundations of Mathematics. 1995 ISBN 0-7923-3484-1
252. A. Wiśniewski, *The Posing of Questions*. Logical Foundations of Erotetic Inferences. 1995 ISBN 0-7923-3637-2
253. J. Peregrin, *Doing Worlds with Words*. Formal Semantics without Formal Metaphysics. 1995 ISBN 0-7923-3742-5
254. I.A. Kieseppä, *Truthlikeness for Multidimensional, Quantitative Cognitive Problems*. 1996 ISBN 0-7923-4005-1
255. P. Hugly and C. Sayward: *Intensionality and Truth*. An Essay on the Philosophy of A.N. Prior. 1996 ISBN 0-7923-4119-8
256. L. Hankinson Nelson and J. Nelson (eds.): *Feminism, Science, and the Philosophy of Science*. 1997 ISBN 0-7923-4162-7
257. P.I. Bystrov and V.N. Sadovsky (eds.): *Philosophical Logic and Logical Philosophy*. Essays in Honour of Vladimir A. Smirnov. 1996 ISBN 0-7923-4270-4
258. Å.E. Andersson and N-E. Sahlin (eds.): *The Complexity of Creativity*. 1996 ISBN 0-7923-4346-8
259. M.L. Dalla Chiara, K. Doets, D. Mundici and J. van Benthem (eds.): *Logic and Scientific Methods*. Volume One of the Tenth International Congress of Logic, Methodology and Philosophy of Science, Florence, August 1995. 1997 ISBN 0-7923-4383-2
260. M.L. Dalla Chiara, K. Doets, D. Mundici and J. van Benthem (eds.): *Structures and Norms in Science*. Volume Two of the Tenth International Congress of Logic, Methodology and Philosophy of Science, Florence, August 1995. 1997 ISBN 0-7923-4384-0
Set ISBN (Vols 259 + 260) 0-7923-4385-9
261. A. Chakrabarti: *Denying Existence*. The Logic, Epistemology and Pragmatics of Negative Existentials and Fictional Discourse. 1997 ISBN 0-7923-4388-3
262. A. Biletzki: *Talking Wolves*. Thomas Hobbes on the Language of Politics and the Politics of Language. 1997 ISBN 0-7923-4425-1
263. D. Nute (ed.): *Defeasible Deontic Logic*. 1997 ISBN 0-7923-4630-0
264. U. Meixner: *Axiomatic Formal Ontology*. 1997 ISBN 0-7923-4747-X
265. I. Brinck: *The Indexical 'I'*. The First Person in Thought and Language. 1997 ISBN 0-7923-4741-2
266. G. Hölmström-Hintikka and R. Tuomela (eds.): *Contemporary Action Theory*. Volume 1: Individual Action. 1997 ISBN 0-7923-4753-6; Set: 0-7923-4754-4
267. G. Hölmström-Hintikka and R. Tuomela (eds.): *Contemporary Action Theory*. Volume 2: Social Action. 1997 ISBN 0-7923-4752-8; Set: 0-7923-4754-4
268. B.-C. Park: *Phenomenological Aspects of Wittgenstein's Philosophy*. 1998 ISBN 0-7923-4813-3

MIX
Papier aus verantwortungsvollen Quellen
Paper from responsible sources
FSC® C105338

If you have any concerns about our products,
you can contact us on
ProductSafety@springernature.com

In case Publisher is established outside the EU,
the EU authorized representative is:
Springer Nature Customer Service Center GmbH
Europaplatz 3, 69115 Heidelberg, Germany

Printed by Libri Plureos GmbH
in Hamburg, Germany